carpentry
in
residential
construction

carpentry in residential construction

SECOND EDITION

Stanley Badzinski, Jr.

Milwaukee Area Technical College
Milwaukee, Wisconsin

Prentice-Hall, Inc., Englewood Cliffs, New Jersey 07632

Library of Congress Cataloging in Publication Data

BADZINSKI, STANLEY.
    Carpentry in residential construction.

    Bibliography: p.
    Includes index.
    1. Carpentry. 2. Building, Wooden. 3. House
construction. I. Title.
TH5604.B23 1980   694   80-13094
ISBN 0-13-115238-6

Editorial/production supervision
  and interior design by Karen Skrable
Cover design by Jorge Hernandez
Manufacturing buyer: Anthony Caruso

Printed in the United States of America

10 9 8 7 6 5 4 3 2

Prentice-Hall International, Inc., *London*
Prentice-Hall of Australia Pty. Limited, *Sydney*
Prentice-Hall of Canada, Ltd., *Toronto*
Prentice-Hall of India Private Limited, *New Delhi*
Prentice-Hall of Japan, Inc., *Tokyo*
Prentice-Hall of Southeast Asia Pte. Ltd., *Singapore*
Whitehall Books Limited, *Wellington, New Zealand*

To the Student:

Carpentry, like all skilled trades, requires study and practice to attain proficiency. Do not minimize the importance of study if you wish to be a truly skilled carpenter.

As a young man contemplating a vocation my father offered the following advice which I pass on to you: ". . . if you want to be a carpenter, study hard, and be a good one."

STANLEY BADZINSKI, JR.

contents

6   UNEQUAL PITCH INTERSECTING ROOFS   173

7   EXTERIOR TRIM AND EXTERIOR SIDING   195

8   BUILDING SERVICES   241

preface

Carpentry in Residential Construction is written for the carpenter apprentice engaged in home and apartment building and for students enrolled in architectural technology and construction technology at two-year technical institutes.

Information is presented on the principles of good foundation work and on the importance of a good foundation to carpentry work.

The various types of frame construction are discussed and the effect of lumber shrinkage on the building frame is explored along with a discussion of the advantages and disadvantages of each type of frame construction.

In the chapter on floor construction the sizing and installation of beams, columns, and floor joists are discussed. Methods of framing floor openings for fireplaces, chimneys, and stairs are illustrated along with methods of determining the size of headers and trimmers. Also included is a discussion on the installation of bridging and subflooring.

Wall construction is covered from layout to assembly with details given on determining stud length, layout of door and window openings, bracing, and sheathing.

Roof construction is covered in detail with many drawings

showing the layout of all types of rafters for equal and unequal pitch intersecting roofs. A section is included on the installation of roof sheathing and on the types of roofing which carpenters may install.

The chapter on exterior trim and sidings presents information on the various types of cornices, exterior trim, and sidings commonly applied by carpenters as well as various types of door and window frames.

A chapter is included on building services—plumbing, heating, and electrical installations—to acquaint the student with these installations and to help him understand how to avoid placing framing members in areas where the other tradesmen will need room.

Insulation in many areas is installed by specialists. However, the carpenter is also called upon to do insulation work, and a discussion on insulation is presented to acquaint him with the principles of insulation.

In residential work there are many types of materials applied to the walls and ceilings. While the carpenter does not install all of these he must work around them at some time during the course of construction. Therefore, the most common wall materials—lath and plaster, gypsum drywall, prefinished paneling, and other wallboards are discussed.

Following the normal sequence of construction, the next item to be installed is finish flooring. Hardwood strip flooring, parquet wood block flooring, and underlayment for resilient flooring are discussed in detail to provide the student with adequate knowledge of the material and installation procedures.

The chapter on interior trim covers the installation of door jambs, casings and stops, followed by window trim, base trim, closet shelving, cabinet installation, and a discussion on stairways. The hand of doors and the installation of interior doors are illustrated and discussed in detail to enable the student to approach the task of door installation with confidence.

I wish to take this opportunity to thank the many companies, associations, and individuals who contributed pictures and technical material included in this book. A special thank you is given to my wife, Alice, for the many hours of typing and proofreading which were necessary to complete this manuscript.

carpentry
in
residential
construction

footings and foundations

1

Foundations and footings are not thought of as being carpentry work, and indeed they are not built by the carpenter. However, the carpenter should have knowledge of what a good footing is and what it does.

The footing supports the entire structure. It must be of proper size and shape or it will allow the building to settle. This settling is usually uneven and causes cracking in the foundation walls, sagging floors, unsightly cracks in plastered walls, and sticking doors and windows.

When floors sag, windows stick, and doors will not close properly, the carpenter may be accused of doing poor work when the real cause of the problem is a faulty foundation. Therefore it is important for the carpenter and anyone else concerned with residential construction to understand the importance of a good foundation.

SOIL BEARING CAPACITIES

The ability of a soil to carry a load is referred to as its supporting capability. Different soils are capable of carrying greater or smaller loads per square foot and as a result have greater or smaller bearing capacities.

Safe bearing capacities for the different types of soils are usually given in pounds per square foot or tons per square foot. In some cases bearing capacities are established by laboratory testing, but often they are established through common practice and experience over a number of years.

Some of the factors which affect a soil's ability to carry loads are its composition, its depth, the amount of water it contains, its location, and its confinement. If a soil contains decaying organic material it will be a relatively poor bearing soil because natural settling will take place. A relatively thin layer of soil with good bearing capacity which lies on a poor bearing soil cannot deliver full support. Therefore the depth of the bearing soil is important.

The bearing capacity of some soil is more affected by the presence of water than others. This factor must be considered when establishing the load-bearing capacity of the soil. A building placed on the side of a hill may have a tendency to slide when soil is moved by water. Because some soils have a greater tendency to move than others, especially when in thin layers, the stability of bearing soil must also be considered. Sand, especially, has a tendency to move due to wind and water unless it is confined or stabilized. Before any construction is begun on doubtful or unstable soil a competent engineer should be consulted.

Some generally accepted safe bearing capacities for soils commonly encountered are given in Table 1-1. It should be

Table 1-1. Bearing capacities for common types of soil

TYPE OF SOIL	BEARING CAPACITY TONS PER SQ. FT.
Rock	
Granite	30
Sandstone	10–20
Limestone	10–20
Hardpan	4
Gravel	4–6
Sand (dry)	2
Sand (wet)	1
Clay (dry)	2–4
Clay (wet)	1–2
Loam	1–4
Loose fill	unstable

noted that these bearing capacities can be varied because of local experience or laboratory analysis.

Rock may be of several varieties. Granite is an igneous rock and is usually strong, durable, and nonporous. It is practically insoluble and makes an excellent bearing material. Sandstone and limestone are sedimentary rocks. There are many types of sandstone. The material which cements the particles of sand together greatly affects its bearing capacity. Although limestone is slightly soluble in water, this characteristic does not affect its bearing capacity unless it is constantly subjected to the action of running water.

Hardpan is a soil in which the particles have been thoroughly cemented together to form a rocklike material which will not soften when wet. True hardpan will prevent the downward flow of water into the soil.

Gravel is coarse sandy soil which contains broken stone.

Sand is a loose granular material. It may contain fine, medium, or coarse particles and may be mixed with varying amounts of other soils. Its bearing capacity may be greatly affected by water flowing through it.

Clay is a fine-textured soil which breaks into clods or lumps that are very hard when dry. It is a stiff bearing soil when dry but becomes soft when saturated with water. Because of this characteristic it is generally considered an unstable bearing soil.

Loam is a soil which may contain sand, clay, and organic materials. It is often referred to as a top soil. When it contains a large amount of organic material it becomes an unstable bearing soil.

Loose fill may contain any or all of the types of soil mentioned. Because it is loose it does not make a good bearing soil. Before building on fill it is necessary to compact it either by mechanical means or by allowing it to settle naturally. Natural settling may take two or more years to accomplish. When building on a filled area uneven settling of the building will sometimes occur. This is often caused by placing part of the building over filled soil and part on virgin soil. Because fill is unstable, special precautions should always be taken when it is used to support a building. These precautions include making a laboratory analysis of the soil and designing the footings to carry the building safely. If a laboratory analysis cannot be made the footings should be made one and one-half to two times larger than normal for virgin soil.

WALL FOOTINGS

Footings for any permanent building should be placed below the frost line. The frost line is the average depth to which frost penetrates the soil in the winter. It is generally established at about 48 inches below the existing grade but may be at greater or lesser depths, depending on the amount of water in the soil and the low winter temperature.

Wall footings support the loads from floors, walls, ceilings, and roofs carried to them by the foundation wall. These loads include the weight of the building framework and the live loads imposed by furniture and building occupants.

The wall footings should be formed with the sides vertical and the bottom flat and horizontal. If this is not done the resulting footings may be of irregular shape and improper size (see Fig. 1-1).

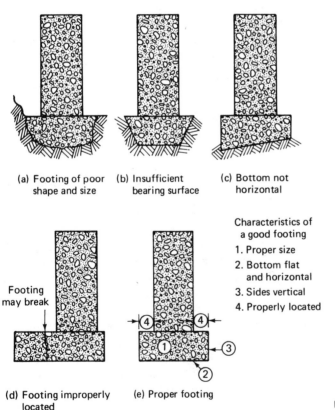

(a) Footing of poor shape and size

(b) Insufficient bearing surface

(c) Bottom not horizontal

Footing may break

(d) Footing improperly located

(e) Proper footing

Characteristics of a good footing
1. Proper size
2. Bottom flat and horizontal
3. Sides vertical
4. Properly located

Fig. 1-1   Wall Footings

Footings supporting solid concrete foundation walls may require a keyway to aid in bonding the wall to the footing. If a keyway is required by code it should be of such size as would not weaken the footing. It is interesting to note that when concrete blocks are used to build the foundation wall the keyway is not needed. Therefore one might also question the need for the keyway when the wall is made of solid concrete.

The size of wall footings is generally established by local building codes; sometimes it may be established by general practice in the locality. Wall footings of plain concrete (concrete without reinforcing steel) are usually 6″ to 8″ thick and formed with 2″-thick lumber staked into place at intervals to maintain size and location (see Fig. 1–2). When concrete footings are formed by the sides of a trench, care must be taken to remove all loose soil, which has a tendency to fall from the sides of the excavation.

A general rule for the width of a wall footing states that the footing should be twice the width of the foundation wall

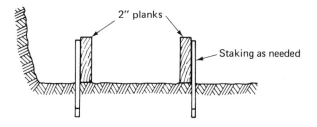

Wall footing form

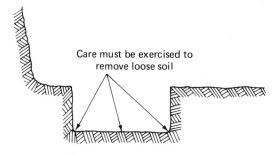

Footing formed by trench   Fig. 1–2   Wall Footing Forms

but should not project more than 6″ beyond either side of the wall unless it is reinforced.

COLUMN FOOTINGS

Footings which support columns may carry a load 10 times as great per square foot as wall footings in the same building. Because of this concentrated loading, care should be taken when determining the size of these footings as well as in constructing them.

The general rule for plain concrete column footings states that the thickness of the footing should be equal to one-half its longest side. Therefore the most economical shape for a column footing is a square. The size of the footing can be determined after the load on the column is known. This load includes the live load and the weight of floor and wall construction supported by the beam section carried by the column (see Fig. 3–8). It is also necessary to know the allowable load for the bearing soil.

EXAMPLE

Total load carried by the column = 16,000 lb

Bearing soil (dry clay) allowable load = 4000 lb/sq ft

Footing area = 16,000 ÷ 4000 = 4 sq ft

Footing size = $\sqrt{4}$ = 2 ft × 2 ft

FOUNDATION WALLS

All foundation walls for permanent buildings should be supported on footings of sufficient size placed below the frost line. This is especially important for buildings with masonry veneer walls, because the smallest amount of uneven movement in the footing will cause unsightly cracking in the mortar joints. Wood framework has the ability to bend and move somewhat in the event of unequal settling. However, the plaster or wallboard attached to the framework will show the effects of movement, so the importance of adequate footings should not be minimized.

Concrete Block Foundation

Concrete block foundation walls may be 8″, 10″, or 12″ thick. An 8″-thick wall is usually adequate for a building of all-frame construction (one with no masonry veneer). When a masonry veneer is used a 10″- or 12″-thick wall is built to provide a 4″ ledge for the veneer and sufficient support for the wood framework (see Fig. 1-3).

The concrete blocks which are placed below grade are given a ½″-thick application of cement mortar backplaster to improve the blocks' resistance to ground water. Additional waterproofing is obtained by applying a coating of asphaltic waterproofing over the cement backplaster. This waterproofing material may be painted on with large brushes or sprayed on by a contractor who specializes in basement waterproofing.

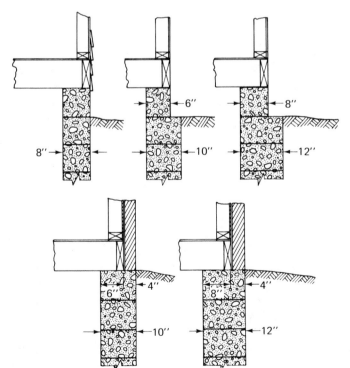

Fig. 1-3 Concrete Block Foundation Walls

When soil conditions are very wet, a heavy-duty waterproofing material made of asphaltic synthetic rubber may be applied over the backplaster. This material is applied with a trowel or special spray equipment to a thickness of approximately $\frac{3}{16}''$ and creates a wall which is highly resistant to water infiltration.

Solid Concrete Foundation

Solid concrete foundation walls for residential constructing are poured in thicknesses from 6″ to 10″. A 4″-wide brick ledge may be formed in the 8″- and 10″-thick walls. Most of these walls do not contain any reinforcing steel.

Formwork for walls of this type is built by carpenters in most areas. This formwork may be of a patented type of plywood and steel or metal panels, or it may be job built using lumber and plywood for the form members (see pp. 22-24).

Solid concrete walls are often waterproofed by applying a coat of asphalt to the outside of the wall below the finish grade. This coating may be applied by brush or spray. It is usually more economically applied by spraying.

When soil conditions are very wet, a heavy-duty waterproofing material made of asphaltic synthetic rubber may be applied. This material is applied with a trowel or special spray equipment to a thickness of approximately $\frac{3}{16}''$ and creates a wall which is highly resistant to water infiltration.

DRAIN TILE

Ground water must be carried away from footings and foundation walls. This water is carried away by installing drain tile around the perimeter of the foundation and leading the water to a point away from the foundation. The drain tile are either 3″ or 4″ in inside diameter and 12″ long. They may be made of clay or concrete.

A new type of "drain tile" made of plastic has been developed and is in use in some areas. It comes in 100 ft rolls and is perforated to allow the water to enter.

In most cases the drain tile is placed on top of the footing as shown in Fig. 1-4 and covered with a 12″ layer of stone. The

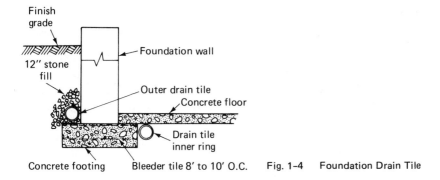

Finish grade

12″ stone fill

Foundation wall

Outer drain tile

Concrete floor

Drain tile inner ring

Concrete footing   Bleeder tile 8′ to 10′ O.C.   Fig. 1–4   Foundation Drain Tile

stone is used to prevent loose soil from entering the tile and clogging the system. Bleeders are placed at 8′ and 10′ intervals to carry the water from the outside ring of tile to an inner ring. The inner ring carries the water to a central collecting point from which it can be pumped away from the building.

PLOT PLANS

The plot plan is a plan of the land on which a proposed structure will be built. The plot plan may be drawn by any person who has the necessary information. Often the plot plan is drawn by the architect, who works from the owner's sketch or from a survey of plot.

The survey of plot is a plot plan drawn by a surveyor after he has checked the legal description of the property and made the survey. Plot plans and the survey of plot show the outline of the property, the locations and names of streets, sidewalks, the outline of existing structures on the property and on adjacent properties, grade elevations, and sometimes the location of trees and other information.

In addition to the preceding information the survey of plot will contain the owner's name, the legal description, and the surveyor's certification.

The surveyor will drive wood stakes to locate the corners of the proposed building. However, there are times when it is necessary to locate a building without the help of the surveyor. Then the building may be staked out by measuring from known points and lines, along with the use of the 3–4–5 rule.

STAKING OUT

The 3-4-5 rule can be used to lay out a square corner because the numbers 3, 4, and 5 represent the sides of a right triangle. Any multiple of 3, 4, and 5 may be used (see Table 1-2). It is best to use the largest multiple possible for the size of building being staked out.

To stake out a building the following steps may be followed (see Fig. 1-5):

1. Locate lot lines.

2. Locate one corner of building with stake.

3. Locate second corner of building by measuring from lot lines and first stake. Drive second stake.

Table 1-2. Multiples of 3-4-5

a	b	c	a	b	c
3	4	5	18	24	30
6	8	10	21	28	35
9	12	15	24	32	40
12	16	20	27	36	45
15	20	25	30	40	50

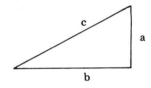

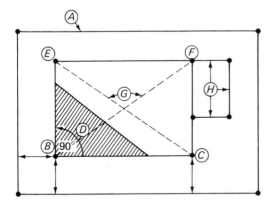

- Ⓐ   Locate lot lines
- Ⓑ   Locate first corner
- Ⓒ   Locate second corner
- Ⓓ   Layout square corner (3-4-5 rule)
- Ⓔ   Locate third corner
- Ⓕ   Locate fourth corner
- Ⓖ   Check layout for squareness
- Ⓗ   Make additional layouts

Fig. 1-5   Staking Building Location

4. Lay out square corner using 3–4–5 rule.

5. Locate third corner.

6. Locate fourth corner by measuring from established corners.

7. Measure diagonals to check squareness of overall layout.

8. Make additional layouts as required.

Batterboards

To maintain the location of building lines while the basement is being excavated and to provide established reference points batterboards are sometimes used. These batterboards are L-shaped structures located far enough away from the building so as not to interfere with the excavation. They are well braced and have the building lines established on them (see Fig. 1–6).

After the excavation is completed lines are stretched between batterboards, and plumb bobs are dropped from the in-

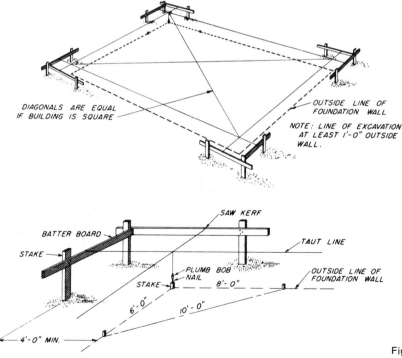

DIAGONALS ARE EQUAL
IF BUILDING IS SQUARE

OUTSIDE LINE OF
FOUNDATION WALL

NOTE: LINE OF EXCAVATION
AT LEAST 1'-0" OUTSIDE
WALL.

SAW KERF

BATTER BOARD

STAKE

TAUT LINE

PLUMB BOB
NAIL

STAKE

8'-0"

OUTSIDE LINE OF
FOUNDATION WALL

6'-0"

10'-0"

4'-0" MIN.

Fig. 1–6   Batter Boards

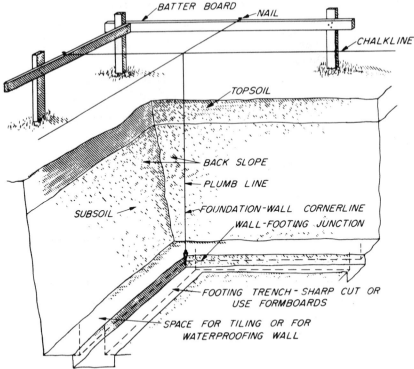

BATTER BOARD
NAIL
CHALKLINE
TOPSOIL
BACK SLOPE
PLUMB LINE
SUBSOIL
FOUNDATION-WALL CORNERLINE
WALL-FOOTING JUNCTION
FOOTING TRENCH - SHARP CUT OR USE FORMBOARDS
SPACE FOR TILING OR FOR WATERPROOFING WALL

Fig. 1-6   Continued

tersecting lines to locate the building lines at the floor of the excavation.

GRADE ELEVATIONS

The heights of floors, footings, and so on are often given as a grade in feet and inches or feet and hundredths of a foot. The heights or grades are established from a known reference point, which is often referred to as a benchmark or datum point. These reference points may be established benchmarks based on sea level, or some stable object such as the intersection of concrete streets or sidewalks, a concrete slab, or a bolt on a fire hydrant. When grade elevations are given on a set of plans the benchmark will also be stated.

A transit or builder's level and a surveyor's target rod are needed to establish grade elevations.

THE LEVEL-TRANSIT

The level-transit is a precision instrument which is sturdily built but requires caution and care in handling. It is, therefore, necessary to avoid dropping the unit, bumping it against objects, or handling it in any careless manner.

Level-transits made by different manufacturers have different features, but all should be kept clean, dry, and properly adjusted. When not in use they should be stored in the carrying case in accordance with the manufacturer's instructions. If the transit should need adjustment because of careless handling or an accident, it is best to return the unit to a qualified repair station where it can be repaired under controlled conditions.

Transit Nomenclature

Before attempting to use a transit the prospective user should become familiar with the features of the instrument. He should also familiarize himself with its operation. Transits made by different manufacturers will have different operating features, and as each different make or type of transit is encountered it should be studied carefully before attempting its use.

The level-transit shown in Figs. 1-7 and 1-8 is one type commonly used in residential construction. If you study the illustration you will be able to familiarize yourself with features common to most transits.

Telescope   The optical instrument through which sighting and reading is done. The magnifying power of the telescope in Figs. 1-7 is 26 times.

Objective Lens   The lens at the large end of the telescope.

Focusing Knob   The knob on top of the telescope used to focus the scope on objects at different distances.

Eyepiece Cap   The cap over the eyepiece which is rotated to focus the crosshairs.

Crosshair Adjusting Screw   The screws used to adjust the crosshairs. Crosshairs should be adjusted at qualified repair stations.

Standards   The frame which supports the telescope.

Telescope Locking Levers   The levers used to lock the telescope

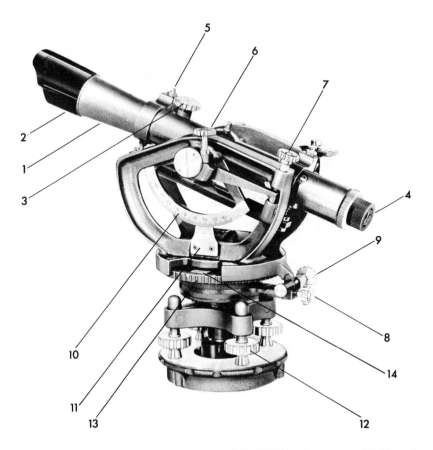

Fig. 1-7   Transit Nomenclature (Courtesy of David White Instruments, Division of Realist, Inc.)

at right angles to the vertical axis of instrument. The levers are unlocked to use the instrument in transit position.

Horizontal Motion Clamp   The screw used to lock horizontal motion of the transit.

Horizontal Motion Tangent Screw   The screw used to make fine horizontal adjustments after locking horizontal motion.

Horizontal Circle   The circle at the base of the instrument which is graduated in degrees.

Horizontal Circle Vernier   The scale used for reading fractions of a degree.

Leveling Baseplate   The plate which attaches the transit to the tripod and serves as a base for the leveling screws.

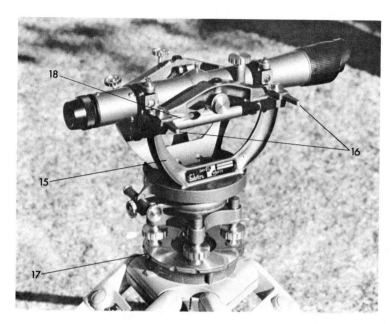

Fig. 1-8   Transit Nomenclature (Continued)

Leveling Screws   The screws which are adjusted to level the instrument.

Vertical Motion Clamp   The screw used to lock vertical motion of the telescope.

Vertical Motion Tangent Screw   The screw used to make fine vertical adjustments after locking the vertical motion clamp.

Vertical Arc   The arc graduated in degrees for measuring upward and downward angles.

Vertical Arc Vernier   The vernier used to read parts of a degree on the vertical arc.

Telescope Level Tube   The level tube attached to the telescope used to level the instrument.

Tripod   The three-legged support on which the transit is installed for use.

Transit Setup

The tripod should be set up with the legs set firmly before the transit is removed from the carrying case. As a general rule the legs should have a spread of about 3 ½′ (see Fig. 1-9), and

Fig. 1-9   Transit on Tripod

the head of the tripod should appear to be level. The legs of the tripod should be fastened securely to the tripod head, and adjustable legs should be tightened to avoid slippage when the transit is placed on the tripod.

Before removing the instrument from the case the horizontal motion clamp should be loosened. Then, grasping the instrument firmly in one hand, the leveling baseplate can be turned free of the carrying case.

Carefully place the instrument on the tripod and fasten it securely by turning the leveling baseplate, but do not overtighten. If the instrument is to be used to measure an angle the plumb bob must be attached to the hook or chain provided at the base of the transit and adjusted until it is just above grade. If it is necessary to move the instrument to bring it directly over the point this must be done before leveling the transit. A small amount of movement can be obtained by loosening two adjacent leveling screws and moving the transit on the leveling baseplate. If greater movement is needed to bring the transit over the point the tripod must be moved.

When the tripod is finally positioned, be sure that each of the points are well into the ground so that there will be no movement. On paved surfaces, where points cannot be driven, be sure that they are positioned to hold securely.

Leveling the Instrument.   The most important operation in preparing the instrument for use is leveling. A poorly leveled instrument will give false readings, and the amount of error will increase as the distance from the transit to the target increases.

The first step in leveling is to be sure that the instrument is locked in place on the baseplate. This is done by checking the tightness of two adjacent leveling screws. When leveling the transit two opposite leveling screws must be turned the same amount at the same time. The direction in which they are turned is such that the thumb on each screw moves in at the same time or moves out at the same time (see Fig. 1-10).

After checking to be sure the instrument is locked in place the actual leveling may begin by turning the telescope so that it is lined up across an opposite pair of leveling screws. The bubble in the level vial is brought to the center of the vial by adjusting the leveling screws over which the telescope was set. The bubble moves in the direction of the left thumb. Therefore, if the bubble must be moved to the right the left thumb must

Fig. 1-10   Leveling the Transit

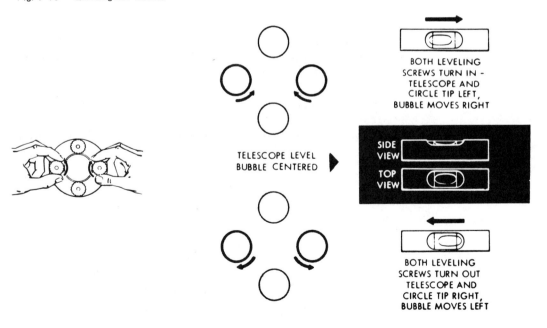

TELESCOPE LEVEL
BUBBLE CENTERED ▶

BOTH LEVELING
SCREWS TURN IN -
TELESCOPE AND
CIRCLE TIP LEFT,
BUBBLE MOVES RIGHT

SIDE VIEW

TOP VIEW

BOTH LEVELING
SCREWS TURN OUT
TELESCOPE AND
CIRCLE TIP RIGHT,
BUBBLE MOVES LEFT

move in to the right, but if the bubble must move to the left the left thumb must move out to the left.

When the bubble is centered in the vial with the telescope over one opposite pair of leveling screws, the telescope is rotated 90° so that it is over the other pair of leveling screws. This pair of leveling screws is adjusted to bring the bubble to the center of the vial. Then the telescope is turned across the first pair of screws and readjusted. After rechecking the level across both pairs of screws it should be possible to rotate the telescope 360° without any change in the position of the bubble.

It should be noted that if the level vial is out of adjustment it is still possible to level the instrument. The procedure is to level the telescope over one pair of leveling screws and then rotate the telescope 180°. The leveling screws are adjusted to bring the bubble back one-half the distance it moved off center. Then the telescope is rotated 90° and the bubble is brought to the same relative position it had before the telescope was turned. After rechecking it should be possible to rotate the telescope in a complete circle with no change in the position of the bubble in the level vial.

MEASURING DIFFERENCES IN GRADE ELEVATION

The transit can be used to measure differences in grade elevations and to transfer grade elevations with the aid of a target rod. In using the transit for this purpose the tripod must be well set and the transit must be leveled accurately. The target rod must be held perfectly vertical. Side-to-side movement can be checked along the vertical crosshair. The smallest reading is obtained when the rod is held vertically. Therefore, if the rod is moved back and forth the smallest reading is the one recorded.

Determining Grade Elevations

Grade elevations for a given job may be related to city datum, but more likely they will be related to a reference point on or near the job which will be arbitrarily assigned an elevation of +100.00'. Elevations of the various points around the job site will be related to this reference point. Higher points will have elevations greater than +100', and lower points will have elevations less than +100'.

In Fig. 1-11 point *A* was established as the benchmark or reference point for the job and assigned an elevation of +100'. Benchmarks must be stable. Therefore they often are the intersections of concrete streets or walks, but any object which will not move can be used as a benchmark.

To determine the elevations at points *B* and *C* the builder's level-transit was set up at a convenient location and carefully leveled. The first reading was taken at the reference point and was found to be 5' 2''. This reading is called backsight. Backsight is the vertical distance from the reference point to the telescope line of sight. When added to the grade elevation, it gives the elevation of the line of sight. This is called the height of the instrument (HI).

The reading at point *B* is 1' 1'', and is called foresight (FS-). It is given a minus sign because it is subtracted from the height of the instrument (HI) to determine the new elevation. In this case 105' 2'' minus 1' 1'' equals 105' 1'' which is the elevation at point *B*.

The reading at point *C* is 7' 4'' and by subtracting it from HI the elevation at point *C* may be determined. In this case 105' 2'' minus 7' 4'' equals 97' 10''. The difference in elevation between points *A* and *C* is determined by subtracting the elevation at *C*, 97' 10'', from the elevation at *A*, 100' 0'', and is found to be 2' 2''.

A simple table for determining grade elevations is given in Fig. 1-12.

When there are great differences in elevations it may be necessary to move the transit a number of times and to take readings in both directions from high and low points to deter-

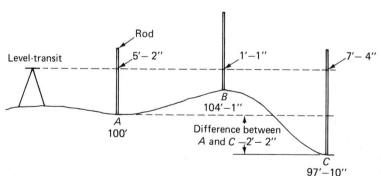

Fig. 1-11   Measuring Differences in Grade

Fig. 1-12   Record of Elevations

Station	BS+	HI	FS-	Elevation	Remarks
Bench Mark (BM)				100′ 00″	
1A	5′ 2″	105′ 2″			
1B		105′ 2″	1′ 1″	104′ 1″	
1C		105′ 2″	7′ 4″	97′ 10″	Difference between A and C - 2′ 2″

mine the various elevations. This problem is seldom encountered in residential construction in so far as the carpenter is concerned.

Transferring Grade Elevations

In transferring grade elevations the builder works a problem similar to that for determining elevations, the only difference being that the reference point or benchmark has been established, and the grade elevations are given on the plan.

To transfer a grade elevation the builder first takes a reading at the established benchmark and then compares it with readings taken at other points around or near the building site. He may install stakes at these points as references and indicate with appropriate markings whether the tops of these stakes are above or below the final grade.

LAYOUT OF BUILDING LINES

To use the transit to lay out building lines one corner of the building must be located with a stake, and a line representing one side of the building must be established. In Fig. 1-13 the transit is set up at stake A, and with the aid of a plumb bob is centered over the nail in the stake which represents the exact

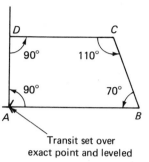

Transit set over
exact point and leveled

Fig. 1-13   Turning an Angle

corner. The transit is then leveled in the usual manner. After leveling the telescope locking levers are opened and a sighting is taken on the established line near point D. To aid in sighting point D the horizontal motion may be locked and the horizontal tangent screw adjusted to sight point D accurately.

With the horizontal motion locked and the vertical crosshair set on point D the horizonatl circle may be turned to 0°. Now the horizontal motion may be unlocked and the transit turned 90° toward point B. After turning the instrument the horizontal motion is again locked. With the telescope in transit position stake B is located at the proper distance from stake A. A nail is driven in the top of stake B to locate the exact corner.

After checking the layout the transit is moved to point B and located and leveled in the same manner as at point A. Following leveling a sighting is taken on point A. The horizontal motion is locked, and the horizontal circle is set to 0°. Then the horizontal motion is unclamped, and the transit is rotated 70°. Stake C is then located in the same manner as was stake B.

To continue the layout the transit is moved to stake C and leveled in the usual manner. The vertical crosshair of the telescope is aligned on stake B, and the horizontal motion is again clamped. After the horizontal circle is set to 0° the horizontal motion is unclamped, and the transit is rotated 110°. Sighting through the transit the vertical crosshair should fall on stake D.

Should any serious errors exist after completing the layout a check of dimensions should first be made and, if necessary, the entire layout should be redone to reconcile the errors.

READING THE VERNIER SCALE

All level transits are equipped with a vernier scale which permits taking readings in parts of a degree. The vernier scale and the obtainable accuracy will vary among different manufacturers, but the procedure in reading the scale is similar for all types.

The vernier in Fig. 1-14 divides each degree into 12 equal parts of 5 min each. While the residential carpenter will seldom need to read angles in parts of a degree, he should be aware of vernier-scale use. The vernier illustrated is marked 60 in the center, which is also 0. In reading the vernier start at 0 and read up the scale until a mark on the vernier and the horizontal circle coincide. The upper illustration reads 77° 20 min.

In reading up the scale, if no mark coincides when we

reach 30 min on the right, move to the 30 on the left and continue reading the scale from the left to right. The lower illustration in Fig. 1–14 reads 76° 45 min.

BUILDING FORMS FOR BASEMENT WALLS

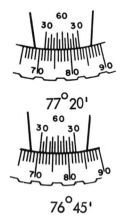

77°20'

76°45'

Fig. 1–14   Reading the Vernier

Formwork for basement walls starts with the building of footing forms. These forms are usually made of 2″-thick lumber of sufficient width, which are held in place with stakes. Footing forms are located at the bottom of the excavation in accordance with the building lines established earlier. The top of the form is set to grade with the use of a builder's level or transit. After one corner of the form is set to grade the entire form is leveled with the aid of a builder's level, and the forms are ready for the concrete.

Plain Wall Forms

Wall forms may be job built using 2 by 4's for plates, studs, and walers (see Fig. 1–15). The form sheathing is usually ⅝″ or ¾″ concrete forming grade plywood. Various types of hardware are available for form ties and spreaders.

Many patented forming systems employing modular panels are available for wall forming. These panels usually consist of a steel frame with plywood sheathing. Some form panel systems are all steel. Each forming system has its own set of hardware which must be used to make the system work efficiently. The procedure followed when erecting forms will vary with the forming system used, but there are many similarities in the erection sequences of the various systems.

The general erection sequence for form panels is summarized as follows:

1. Snap chalk lines on footings to locate outside of foundation walls.
2. Plumb and brace end panel on outside corner.
3. Set remaining panels in place along layout line, and install ties and connecting hardware.
4. Install walers and braces as required.

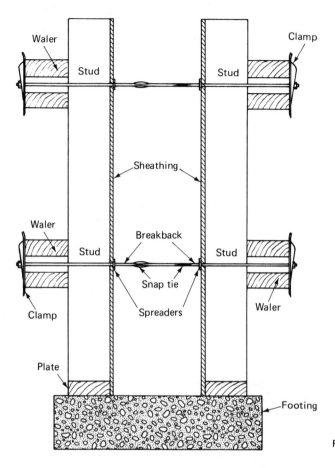

Fig. 1-15   Typical Wall Form

5. Install door bucks and window box-outs before setting interior forms.

6. Set interior form panels.

7. Install walers and bracing as required.

8. Check form alignment.

Wall Forms with Offsets

Walls often require an offset for a brick ledge or a pilaster to strengthen the wall. These offsets require special attention in designing and building the forms. Each forming system has accessories which can be used to advantage when forming offsets, and the various manufacturers' representatives are ready to help

Fig. 1–16   Offsets in Wall Forms (Courtesy Allenform Corporation)

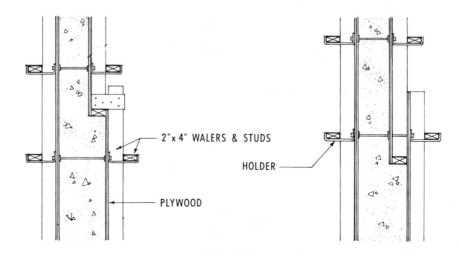

2"x 4" WALERS & STUDS

HOLDER

PLYWOOD

TYPICAL BRICK LEDGE SECTIONS

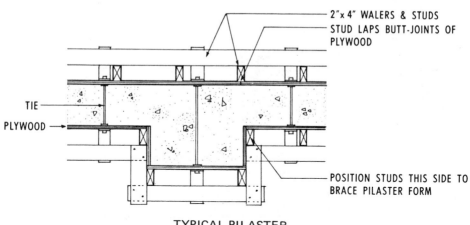

2"x 4" WALERS & STUDS
STUD LAPS BUTT-JOINTS OF
PLYWOOD

TIE

PLYWOOD

POSITION STUDS THIS SIDE TO
BRACE PILASTER FORM

TYPICAL PILASTER

when special problems are encountered. Some typical offsets
are illustrated in Fig. 1–16.

REVIEW QUESTIONS

1. What is a footing?
2. What are some effects of a poor footing?

3. What is soil bearing capacity?

4. What factors affect soil bearing capacity?

5. How is soil bearing capacity determined?

6. Define the bearing soil commonly found in your building area.

7. What is the general rule for width of wall footings?

8. How do loads on column footings compare with loads on wall footings?

9. What is the general rule for column footing thickness?

10. What is frost line?

11. Why must frost line be considered?

12. What materials are commonly used for foundation walls?

13. What is the purpose of drain tile?

14. What is a plot plan? Survey of plot?

15. Outline the procedure you would follow to stake out a building using the 3–4–5 rule.

16. When are batterboards used?

17. What are grade elevations?

18. Outline the procedure you would use to set up and level a transit.

19. List and define the various parts of a concrete wall form.

20. Outline the general procedure for building basement wall forms.

types of frame construction

2

Various types of frame construction have been used in different areas of the country. The type of building frame used was often dependent on the materials available and the climatic conditions. As technology improved, the framing methods were adapted to new materials and methods, but the basic principles of the various types of frame construction remain and must be considered in residential design and construction.

BRACED FRAMING

The braced frame is the oldest type of framing used in the United States. It was the type of framing used by the English in colonial times and was characterized by the use of heavy timbers at the floor lines for beams or girts which supported the floor frames and by timber posts and bracing which supported the beams and girts (see Fig. 2-1). The studs in this frame did not support any floor or roof load but served only as support for the exterior siding and interior plastered walls.

Fig. 2-1   Braced Frame (Courtesy of Delmar Publishers, a Division of Litton Educational Publishing, Inc.)

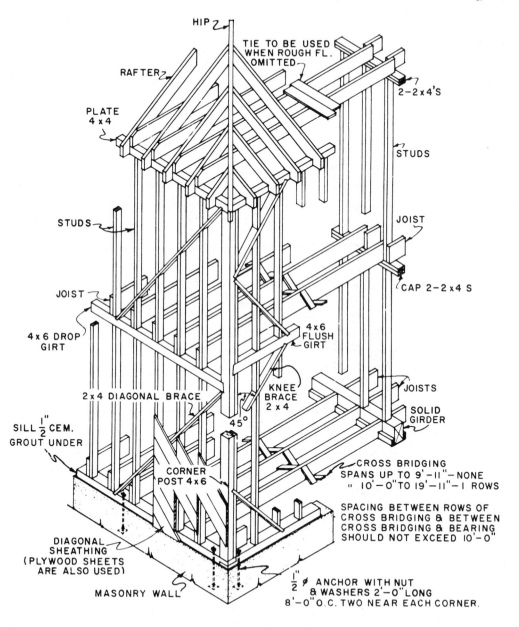

HIP

TIE TO BE USED
WHEN ROUGH FL.
OMITTED

RAFTER

2-2 x 4'S

PLATE
4 x 4

STUDS

STUDS

JOIST

JOIST

CAP 2-2 x 4 S

4 x 6 DROP
GIRT

4 x 6
FLUSH
GIRT

2 x 4 DIAGONAL BRACE

KNEE
BRACE
2 x 4

JOISTS

SOLID
GIRDER

45°

SILL $\frac{1}{2}$" CEM.
GROUT UNDER

CROSS BRIDGING
SPANS UP TO 9'-11"-NONE
"   10'-0"TO 19'-11"-1 ROWS

SPACING BETWEEN ROWS OF
CROSS BRIDGING & BETWEEN
CROSS BRIDGING & BEARING
SHOULD NOT EXCEED 10'-0"

CORNER
POST 4 x 6

DIAGONAL
SHEATHING
(PLYWOOD SHEETS
ARE ALSO USED)

MASONRY WALL

$\frac{1}{2}$" ∅ ANCHOR WITH NUT
& WASHERS 2'-0"LONG
8'-0"O.C. TWO NEAR EACH CORNER.

Where braced framing was used in barn construction the studs were omitted and the sheathing boards were applied vertically.

With the introduction of inexpensive nails, modern tools, and hardware, braced framing has undergone many changes. These changes were also hastened by the reduction in the local supply of heavy timber. Modern braced framing (see Fig. 2–2) is used in some areas of the United States. In this type of framing the heavy timbers have been replaced by 2″-thick framing lumber and the stud walls are used to support the floors and roof.

BALLOON FRAMING

The principal feature of balloon framing is the use of one-piece studs running from the sill at the foundation wall to the plate at the roof line (see Fig. 2–3). This feature makes balloon framing the best type of framing for a two-story brick-veneered building because of the minimum amount of settling due to shrinkage of the framing lumber. A problem of unequal settling due to lumber shrinkage is introduced when the bearing partitions are supported on top of the subfloor. This uneven settling can be overcome by supporting the bearing partition directly on the beam, but in most cases the extra work involved makes this method undesirable. Nevertheless, unequal settling due to lumber shrinkage must be considered and guarded against.

The main advantages of balloon framing are limited settling due to lumber shrinkage, fast erection of two-story walls, and the fact that the installation of pipes in outside walls does not require the cutting of plates.

Some of the disadvantages of balloon framing are the need for 18′ long 2 by 4's for studs, possible erection of bearing partitions before subfloor is placed, and the need for firestops between the studs.

SHRINKAGE OF FRAMING LUMBER

Freshly cut lumber may contain up to 200% water by weight. Some of this water is free water contained within the cell cavities and intercellular spaces of the wood, and some of it is water that has been absorbed into the cell walls in the wood fiber. It

Fig. 2-2 Modern Braced Frame (Courtesy of Delmar Publishers, a Division of Litton Educational Publishing, Inc.)

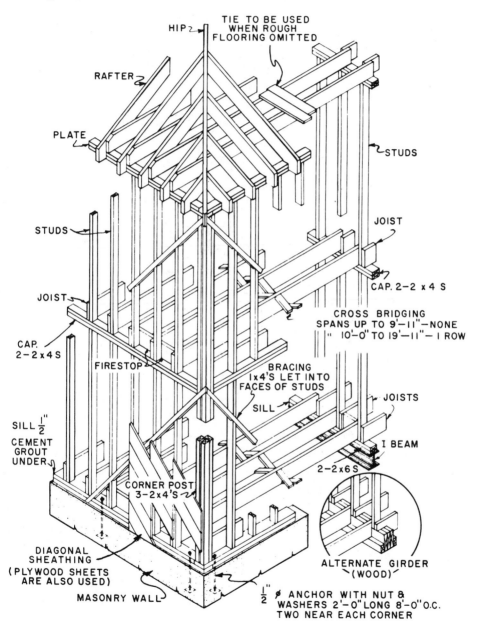

Fig. 2-3   Balloon Frame (Courtesy of Delmar Publishers, a Division of Litton Educational Publishers, Inc.)

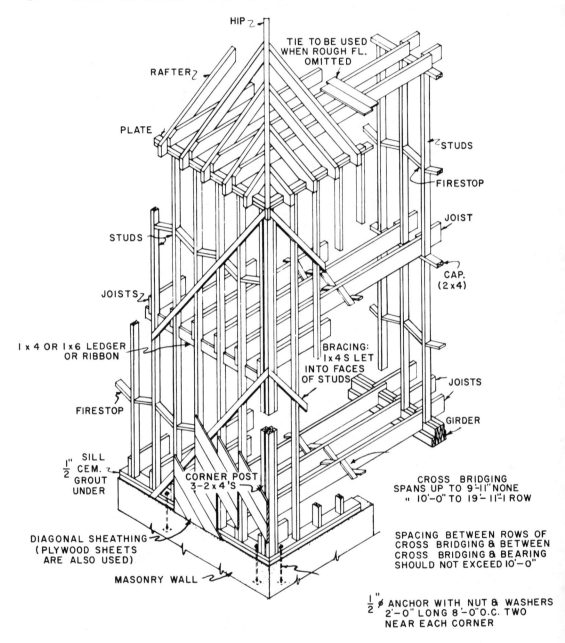

HIP

TIE TO BE USED
WHEN ROUGH FL.
OMITTED

RAFTER

PLATE

STUDS

FIRESTOP

JOIST

STUDS

CAP.
(2 x 4)

JOISTS

1 x 4 OR 1 x 6 LEDGER
OR RIBBON

BRACING:
1 x 4 S LET
INTO FACES
OF STUDS

JOISTS

FIRESTOP

GIRDER

$\frac{1}{2}$" SILL
CEM.
GROUT
UNDER

CORNER POST
3 - 2 x 4's

CROSS BRIDGING
SPANS UP TO 9'-11" NONE
" 10'-0" TO 19'-11"-1 ROW

DIAGONAL SHEATHING
(PLYWOOD SHEETS
ARE ALSO USED)

MASONRY WALL

SPACING BETWEEN ROWS OF
CROSS BRIDGING & BETWEEN
CROSS BRIDGING & BEARING
SHOULD NOT EXCEED 10'-0"

$\frac{1}{2}$" ϕ ANCHOR WITH NUT & WASHERS
2'-0" LONG 8'-0" O.C. TWO
NEAR EACH CORNER

is the water absorbed into the cell walls that causes the wood to swell. When this water leaves the cell walls the wood starts to shrink. Therefore no shrinkage takes place until the moisture content drops below the fiber saturation point, which is about 30% for most species of lumber. As the moisture content drops below 30% the lumber starts to shrink and continues to shrink as moisture is removed until it has reached a point of equilibrium with the surrounding air. This shrinkage occurs in three directions: along the grain or length, radially or across the annual rings, and tangent to the annual rings (see Fig. 2-4).

The shrinkage along the length of a piece of lumber is minimal and amounts to only 0.1–0.3% of the green dimension, and for practical purposes is not considered in frame construction. The greatest amount of shrinkage occurs tangent to the annual rings.

In the radial direction the lumber will shrink about one-half of the amount it shrinks in the direction along the annual rings. This comparative shrinkage must be considered in all types of frame construction.

Properly seasoned lumber reduces the number of problems which occur after a building is completed. It is stronger and more decay resistant. Framing lumber should have a moisture content of 15–19%.

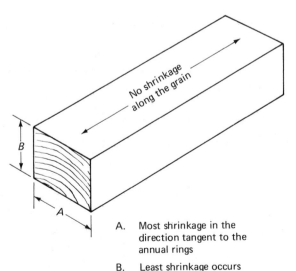

A. Most shrinkage in the direction tangent to the annual rings

B. Least shrinkage occurs radially

Fig. 2-4   Shrinkage of Framing Lumber

PLATFORM FRAMING

The principal features of platform framing are the use of a floor frame or deck covering the entire foundation and the use of walls one story high (see Fig. 2–5). When platform framing is used for buildings of two stories a second platform is built on top of the first-floor walls and walls for the second floor are erected on the second-floor platform. Platform framing is also known as Western framing.

With platform framing the amount of horizontal lumber is equal throughout the building (see Fig. 2–6). Therefore any settling due to shrinkage of the framing lumber is equal throughout the building and is of no consequence unless the building will have a masonry veneer. When a masonry veneer is used precautions must be taken to allow for downward movement of the window frames at the window sill line. This movement occurs when the framing lumber shrinks, causing the entire building frame to settle. The masonry, of course, does not settle but remains in place and forms a barrier to the downward movement of window frames.

Some of the main advantages of platform or Western framing are that platform framing is easy to erect, the platform provides a safe work area during wall erection, all partitions settle the same amount as a result of lumber shrinkage (and therefore the settling is unnoticed), the placing of floor boards is easy because walls are erected after the floor is completed. Also, fire-stops are not needed because the top plates of each wall story provide firestopping automatically.

The main disadvantage of platform framing is the large amount of settling due to lumber shrinkage, which makes it unsuitable for two-story masonry-veneered buildings.

Balloon and platform framing can be combined to gain the advantages of little or no settling in the outside walls due to framing lumber shrinkage and ease of erection of platform frame bearing partitions. However, serious difficulties can result unless provisions are made to allow for settling due to lumber shrinkage.

The outside walls in balloon framing will not settle. However, the bearing partitions which are placed on top of the subfloor will settle as the lumber dries (see Fig. 2–6). Thus the upper floor and ceiling will slope toward the bearing partition. The amount of slope can vary from as little as $\frac{1}{4}''$ to more than $\frac{1}{2}''$.

Fig. 2-5   Platform Framing (Courtesy of Delmar Publishers, a Division of Litton Educational Publishers, Inc.)

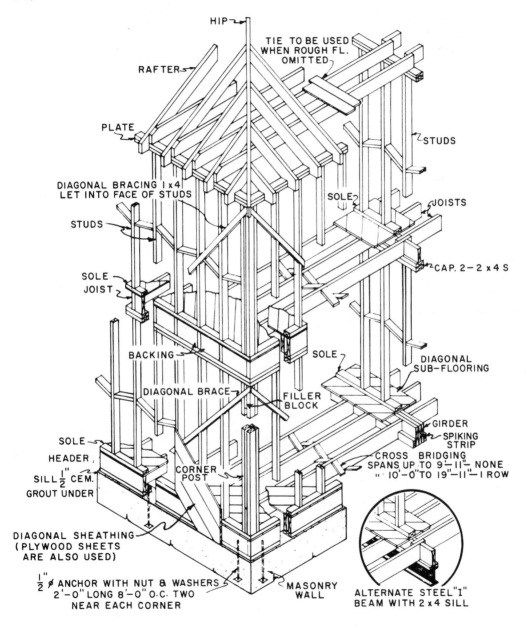

Fig. 2–6   Comparative Settling

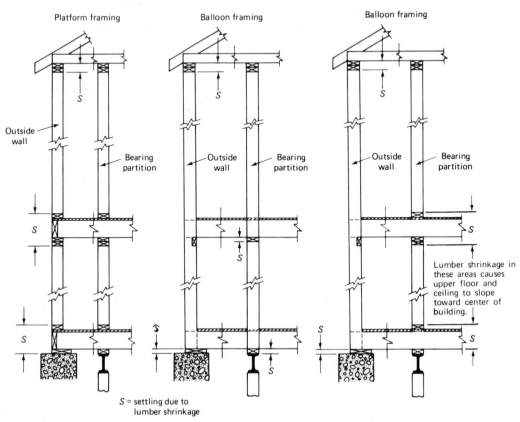

Platform framing

Balloon framing

Balloon framing

Outside
wall

Bearing
partition

Outside
wall

Bearing
partition

Outside
wall

Bearing
partition

Lumber shrinkage in
these areas causes
upper floor and
ceiling to slope
toward center of
building.

S = settling due to
lumber shrinkage

To compensate for this the studs used in the interior partitions should be cut longer. This results in having the upper floor and ceiling sloping upward initially. Therefore sufficient time must be allowed for the lumber to dry out before applying the interior wall material (drywall, plaster, or paneling). Regardless of the approach care must be taken to avoid finishing problems when combining framing methods.

TRI-LEVEL CONSTRUCTION

Many early tri-level homes were built utilizing the platform method of framing, and it was not until difficulties caused by uneven settling due to shrinkage of framing lumber began to

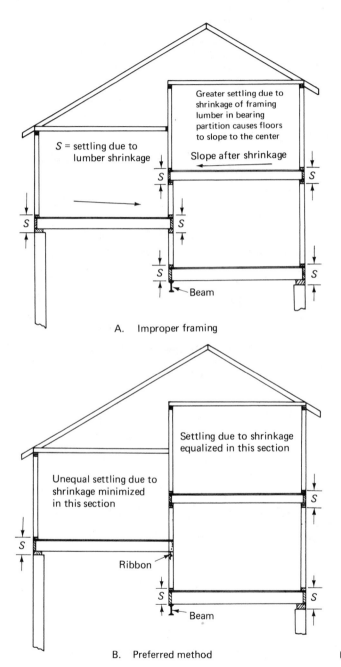

Greater settling due to shrinkage of framing lumber in bearing partition causes floors to slope to the center

S = settling due to lumber shrinkage

Slope after shrinkage

S

S

S

S

S

S

Beam

A. Improper framing

Settling due to shrinkage equalized in this section

Unequal settling due to shrinkage minimized in this section

S

Ribbon

S

S

S

S

Beam

B. Preferred method

Fig. 2-7   Tri-Level Construction

show up in the form of floors out of level that many builders remembered about lumber shrinkage and settling. Because of the different levels of the floor and the manner in which the floors may be supported, tri-level construction usually requires a combination of platform and balloon framing. Figure 2–7 illustrates the proper and improper way to frame on type of tri-level building.

The key to avoiding uneven settling due to lumber shrinkage is to keep the amount of horizontal lumber approximately equal at each level in all bearing partitions.

PLANK-AND-BEAM FRAMING

The number of pieces of framing lumber is reduced by the use of plank-and-beam framing. However, when the framing members are placed farther apart it becomes necessary to use heavier framing lumber (see Fig. 2–8).

Closely spaced floor joists are replaced by beams, usually placed 4′ on center, and 2″ planks or heavy plywood replace the nominal 1″-thick subfloor boards. This system gets its name from the use of beams for joists and planks for flooring.

Beam-type rafters are supported on posts and are spaced at the same intervals as the floor beams. A plank-type roof sheathing acts as the ceiling when the beams are exposed and is covered with rigid insulation to reduce the flow of heat through the roof.

Intermediate studs are placed between the posts to provide a support for the wall sheathing and interior plaster or wallboard.

Plank-and-beam framing reduces the amount of material needed to frame a building. The owner, however, must be willing to accept a sloping ceiling. For this and other reasons related to cost, plank-and-beam framing has not been widely used.

WOOD FRAMING ON CONCRETE SLABS

Many homes are built on concrete slabs when basements are not needed or desired. In warm climates these slabs may be placed over a layer of crushed rock placed on grade. When frost pene-

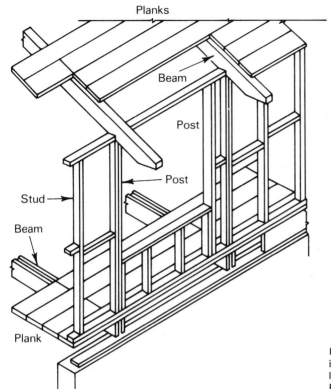

Planks

Beam

Post

Post

Stud

Beam

Plank

Fig. 2-8 Plank and Beam Framing (Courtesy of Delmar Publishers, a Division of Litton Educational Publishers, Inc.)

trates the ground a foundation must be installed with the footings below frost line. The floor slab is placed on a layer of crushed stone and is supported by the foundation wall at the perimeter (see Fig. 2-9). Walls built on this type of slab are usually the same as those used in platform framing.

INNOVATIONS IN FRAME CONSTRUCTION

In an effort to reduce the cost of house construction many new methods of erection and framing are being developed. These new methods strive to reduce the overall cost. Some of these new methods are the box system, the panelized system, rigid frames, and various plywood panel systems, and there are others in the development and experimental stages.

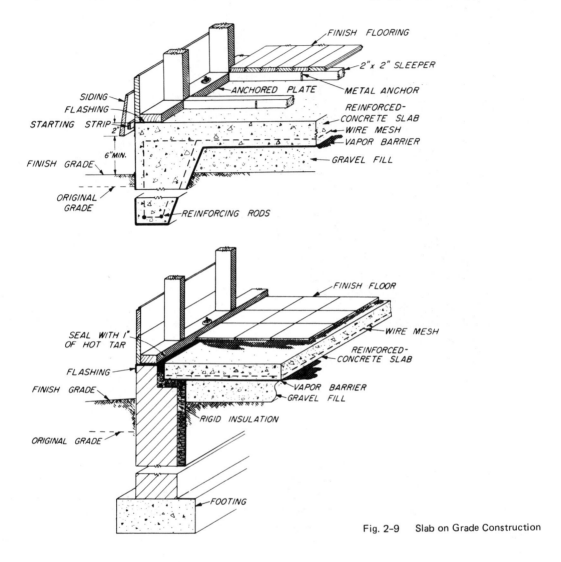

FINISH FLOORING

2" x 2" SLEEPER

ANCHORED PLATE

METAL ANCHOR

SIDING

FLASHING

STARTING STRIP

REINFORCED-
CONCRETE SLAB

WIRE MESH

VAPOR BARRIER

2"

6"MIN.

FINISH GRADE

GRAVEL FILL

ORIGINAL
GRADE

REINFORCING RODS

FINISH FLOOR

SEAL WITH 1"
OF HOT TAR

WIRE MESH

REINFORCED-
CONCRETE SLAB

FLASHING

FINISH GRADE

VAPOR BARRIER

GRAVEL FILL

RIGID INSULATION

ORIGINAL GRADE

FOOTING

Fig. 2-9   Slab on Grade Construction

Box System

The box system employs units or boxes assembled in a factory and installed on the job site in a sequence which produces a home of desirable size. Units or boxes for the system are often built in a more or less conventional manner. Savings realized in building under controlled factory conditions are partly or wholly offset by the cost of handling and transporting the large units.

Panelized System

The panelized system of prefabricating homes utilizes the building of wall sections under factory or shop conditions. These wall sections are shipped to the job site and assembled into a living unit. The panelized system is easier to handle and requires less heavy equipment. Many of the panelized-type systems are constructed in a manner similar to that of platform framing.

Rigid Frames

Rigid frames are made with conventional lumber and are fastened together with plywood gussets. These gussets may be nailed in place, but for added strength glue and nails are used. They may be built on the job site or in a shop. The exterior can be covered with plywood or other common building materials.

Plywood Panel Systems

Various plywood panel systems developed by the American Plywood Association can be used to speed building erection. Many of these panel systems may be built at the job site. This feature eliminates the need for transporting completed units over long distances and can represent a saving.

REVIEW QUESTIONS

1. What happens to framing principles as building methods change?
2. What are the main features of braced framing?
3. Where is braced framing used?
4. What are the main features of balloon framing?
5. How does shrinkage of framing lumber affect building settling?
6. What are the main features of platform framing?

7. What are the disadvantages of using platform framing? Advantages?

8. What are some of the problems encountered in tri-level construction?

9. What are the main features of plank-and-beam framing?

10. When are wood frame walls built on a concrete slab?

11. What is the main reason for developing new methods of framing and erection?

12. What present-day examples of new building methods can you find in your community?

floor construction

3

Floor construction in residential structures usually consists of a wood floor supported by floor joists. The wood floor usually consists of a subfloor and a finish floor. Floor joists are supported by the foundation wall and a central beam or girder. Joist size is governed by the floor load and the joist span.

BEAMS

In house construction a beam may be defined as a heavy framing member which takes the place of an inner foundation wall and supports the inner ends of floor joists. This beam may be made of steel or wood. If it is steel it is usually either an I beam or a small WF (wide flange) beam. Wood beams may be solid timbers, or they may be built up of two or more 2″ planks which have been nailed or bolted together on the job (see Fig. 3-1).

The carpenter is often called upon to decide on the size of beam to be used in a residence. Therefore he should have

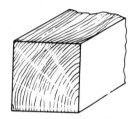

Solid wood beam

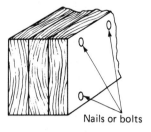

Nails or bolts

Built-up wood beam

Steel beam ("I" beam)

Fig. 3-1   Types of Beams

knowledge of the loads carried by beams and the other factors which affect beam size.

The procedure for determining beam size can be outlined in seven steps:

1. Find the distance between beam supports.
2. Find the beam load width.
3. Find the "total floor load" per square foot carried by the joists and bearing partitions to the beam.
4. Find the load per lineal foot on the beam.
5. Find the total load on the beam.
6. Select the material for the beam.
7. Use beam tables to find the proper size of beam for the material chosen.

Beam Span

The distance between beam supports is known as the beam span and, if not already shown on the plan, must be determined before the beam size can be found. As a general rule the maximum span for beams in residential work is kept down to 8' or 10', to avoid the need for a very large beam required by greater spans and also to cut down on large concentrated loads on the column footings.

Initial beam span is chosen arbitrarily and checked to determine if a satisfactory beam size may be used. The span can then be adjusted to increase or decrease the beam size.

Beam Load Width

Beam load width is the combined distance from the beam to the center of each joist span it supports when the joists are lapped on the beam (see Fig. 3-2). This is true whether the beam is at the center of the building or placed off to one side. Therefore, in a building having one beam with the joists lapped at the beam, the beam load width is equal to one-half the width of the building even though the beam is not centered between the foundation walls.

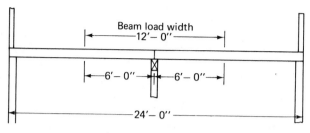

A. Beam centered

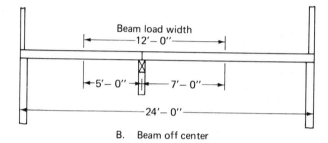

B. Beam off center

Fig. 3–2 Beam Load Width

Continuous Joists

The beam load width for continuous floor joists is greater than that for lapped floor joists having the same span. When a girder is placed at the center of a building and the floor joists are continuous from one foundation wall to the other, the beam load width is equal to five-eighths of the width of the building (see Fig. 3–3). If the beam is moved off center, the beam load width falls somewhere between one-half and five-eighths of the building width. For practical purposes the carpenter should assume that the beam load width for any building with continuous joists is equal to five-eighths of the building width.

Floor Loads

The "total floor load" carried to the beam will depend on the type of building and the number of floors in the building. It is the sum of the various live and dead loads per square foot that are carried to the beams.

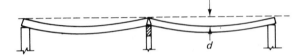

A.   Butted or lapped joists — free end movement

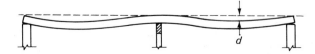

B.   Continuous joist — resistance to bending over
     the beam puts larger proportion of load on beam

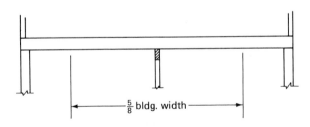

C.   Beam load width is $\frac{5}{8}$ building width when
     the joists are continuous

Fig. 3-3   Beam Load Width – Continuous
Joists

Live loads include all furniture, persons, and other movable loads which are not a permanent part of the structure. Snow on a roof is a live load. Dead loads are those loads which are always present. They include the weight of framing lumber, floors, plaster, tile, plumbing, and so on. The recommended live and dead loads for residential construction are summarized in Table 3-1.

The live load of 40 lb/sq. ft should be used when selecting floor joist size, but when determining beam size for floors over 200 sq. ft, a live load of 30 lb/sq. ft may be used. This reduction is possible because the entire floor will never be fully loaded.

Table 3-1. Recommended loads per square foot of floor and roof construction

TYPE OF LOAD	POUNDS PER SQUARE FOOT
Live load, all floors used as living areas	40
Live load for attic—light storage	20
Dead load, double floor and joists, without plaster	10
Dead load, plastered ceiling and joists	10
Bearing and wall partitions—based on floor area	10
Roof, with asphalt shingles, combined live and dead load	30*

*Roof loads vary greatly with depth of snow, slope, type of roofing, and wind.

The cross section of a building in Fig. 3-4 will serve to illustrate how the various loads in a structure are carried by the beam. Starting at the roof we see that the roof load is carried by the outside walls. Therefore, in this example none of the roof load is carried by the central beam.

When the attic is used for light storage an allowance of 20 lb/sq. ft is made to allow for the weight of the flooring and the live load. Next 10 lb/sq. ft is allowed for the ceiling construction and the plastered ceiling.

The weight of the ceiling and attic floor is carried to the beam by the bearing partitions. Based on floor area an allowance of 10 lb/sq. ft should be made for the weight of bearing partitions and partition walls.

When calculating beam load an allowance of 30 lb/sq. ft is made for the live load on the second floor. The weight of the second-floor construction and plastered ceiling is placed at 20 lb/sq. ft.

From the second-floor level the weight is carried to the first floor and beam by the first-floor bearing partition, and an allowance of 10 lb/sq. ft is made for the first-floor partitions. The live load on the first floor is placed at 30 lb/sq. ft, and the weight of the first-floor construction is placed at 10 lb/sq. ft.

Totaling the various loads, we find that the beam for the building in Fig. 3-4 carries 140 lb/sq. ft of area carried by the beam.

The beam carries the load to the distance of its load width. To find the load carried per lineal foot of beam, the load per square foot carried to the beam is multiplied by the beam load width. For the building in Figs. 3-4 and 3-5 the load per lineal

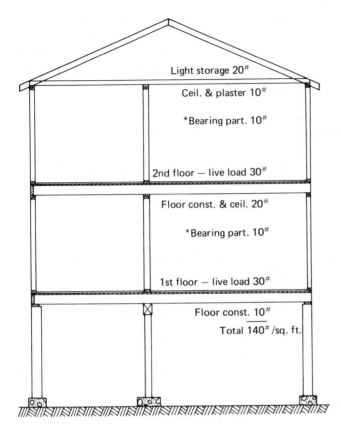

Light storage 20#

Ceil. & plaster 10#

*Bearing part. 10#

2nd floor — live load 30#

Floor const. & ceil. 20#

*Bearing part. 10#

1st floor — live load 30#

Floor const. 10#
_____
Total 140# /sq. ft.

Fig. 3–4   Load Carried by Beam per Square Foot of Floor Area

foot of beam is 140 lb/sq. ft times 14' or 1960 lb per lineal foot of beam.

Total load per beam span can be found by multiplying the load per lineal foot by the span. Therefore 1960 times 8 equals 15,680 lb per beam span for the building in Fig. 3-5.

Selecting Beam Size

The next step in finding beam size is to determine the material to use. Material may be chosen on the basis of availability, desirability of appearance, or owner preference. In residential work the beam is usually either a steel I or WF, or it may be of wood. If it is wood it may be solid or built up. The size of the beam may be determined by consulting tables of beam sizes for uniformly distributed loads. When tables are used care must be taken to use a table which was made for the type and grade of material chosen (see Table 3-2).

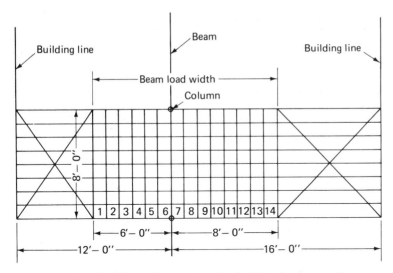

Load per lineal foot of beam = beam load width x load per sq. ft.
LPFB = 14 x 140 = 1960 lbs.
Total beam load = load per lineal foot x beam span
TBL = 1960 x 8 = 15,680 lbs.

Fig. 3-5   Calculating Beam Size

Fig. 3-6 Beam Blocked in Place

Table 3-2. Allowable total load on rectangular wood beams uniformly distributed

Allowable bending stress 1200 psi
Maximum deflection is span over 360 where
   modulus of elasticity is 1,200,000 psi
Horizontal shear 95 psi

Nominal size	2 by 4	4 by 4	2 by 6	3 by 6	4 by 6	6 by 6	2 by 8	3 by 8	4 by 8	6 by 8	8 by 8
Actual width	1.5000	3.5000	1.5000	2.5000	3.5000	5.5000	1.5000	2.5000	3.5000	5.5000	7.2500
Actual depth	3.5000	3.5000	5.5000	5.5000	5.5000	5.5000	7.2500	7.2500	7.2500	7.2500	7.2500
Board feet per foot	0.666	1.333	1.000	1.500	2.000	3.000	1.333	2.000	2.666	4.000	5.333
Area	5.250	12.250	8.250	13.750	19.250	30.250	10.875	18.125	25.375	39.875	52.562
Span is 4′	590	1377	1036	1728	2419	3802	1366	2278	3189	5012	6606
5′	374	874	1034	1724	2414	3794	1364	2273	3183	5002	6594
6′	257	599	995	1659	2323	3652	1361	2269	3177	4992	6581
7′	185	432	740	1234	1727	2715	1358	2264	3170	4982	6568
8′	138	323	561	936	1310	2059	1292	2154	3016	4740	6248
9′			438	730	1022	1607	1021	1702	2383	3746	4938
10′			349	582	815	1282	820	1367	1914	3008	3965
11′			283	472	661	1039	670	1117	1565	2459	3242
12′			232	387	542	853	556	927	1298	2040	2689
13′			192	321	449	706	466	777	1089	1711	2255
14′			160	267	374	588	395	658	921	1448	1909
15′						492	336	561	785	1234	1627
16′						411	288	480	673	1057	1394
17′						344	248	413	578	909	1199
18′						286	213	356	498	783	1033
19′								307	430	675	891
20′								264	370	582	767
21′										500	659
22′										428	564
23′										364	479
24′										306	403
25′											
26′											
27′											
28′											
29′											
30′											

BEAM INSTALLATION

Beams must be installed so that they are properly located, have sufficient bearing on the foundation wall, and are set to proper grade over their entire length.

In residential work the beam is usually set into a pocket in the foundation wall. The location of this pocket is determined by referring to the building plans. When the beam is set into the pocket steel shims are installed under it if necessary to raise the beam to the proper height. The beam is also located in the pocket at the distance from the side wall indicated on the plan. To keep the beam from moving in the pocket, before the floor

Table 3-2. Continued

2 by 10	3 by 10	4 by 10	6 by 10	8 by 10	10 by 10	2 by 12	3 by 12	4 by 12	6 by 12	8 by 12	10 by 12	12 by 12
1.5000	2.5000	3.5000	5.5000	7.2500	9.2500	1.5000	2.5000	3.5000	5.5000	7.2500	9.2500	11.2500
9.2500	9.2500	9.2500	9.2500	9.2500	9.2500	11.2500	11.2500	11.2500	11.2500	11.2500	11.2500	11.2500
1.666	2.500	3.333	5.000	6.666	8.333	2.000	3.000	4.000	6.000	8.000	10.000	12.000
13.875	23.125	32.375	50.875	67.062	85.562	16.875	28.125	39.375	61.875	81.562	104.062	126.562
1744	2906	4069	6394	8429	10754	2121	3535	4949	7777	10251	13080	15908
1740	2901	4061	6382	8413	10733	2116	3528	4939	7762	10232	13054	15877
1737	2895	4053	6369	8396	10713	2112	3521	4930	7747	10212	13029	15846
1733	2889	4045	6357	8380	10692	2108	3514	4920	7732	10192	13004	15815
1730	2884	4037	6345	8364	10671	2104	3507	4910	7717	10172	12978	15785
1727	2878	4030	6332	8347	10650	2100	3500	4901	7702	10152	12953	15754
1677	2795	3913	6150	8108	10344	2096	3494	4891	7687	10133	12928	15723
1416	2360	3305	5193	6846	8734	2092	3487	4882	7672	10113	12903	15692
1180	1968	2755	5707	7282	2059	3432	4806	7553	9957	12703	15450
996	1661	2326	3655	4818	6147	1818	3031	4244	6669	8791	11216	13641
850	1416	1983	3117	4108	5242	1556	2594	3632	5708	7524	9600	11676
731	1218	1705	2680	3533	4508	1344	2241	3137	4930	6499	8292	10085
633	1055	1477	2321	3059	3903	1170	1950	2730	4291	5656	7217	8777
551	918	1286	2021	2664	3399	1025	1708	2391	3758	4954	6321	7688
482	705	1124	1767	2330	2973	902	1504	2106	3310	4363	5566	6770
423	620	987	1551	2045	2609	798	1330	1863	2927	3859	4924	5989
372	546	868	1364	1799	2295	708	1181	1654	2599	3426	4372	5317
327	481	765	1202	1585	2022	631	1052	1473	2314	3051	3893	4735
289	424	674	1060	1397	1783	563	939	1314	2066	2723	3474	4226
254	374	594	934	1232	1571	503	839	1175	1847	2434	3106	3778
224	328	523	822	1084	1383	450	751	1052	1653	2179	2780	3381
	287	459	722	952	1215	403	672	941	1480	1951	2489	3027
	250	402	632	833	1063	361	602	843	1325	1746	2228	2710
	216	350	550	725	926	323	538	754	1185	1562	1993	2424
		303	476	627	801	288	481	673	1058	1395	1780	2165
		259	408	538	686	257	428	600	943	1243	1586	1929
		219	345	455	581	228	380	533	837	1104	1409	1713

is built, temporary blocking is cut to fit between the beam and foundation wall (see Fig. 3-6).

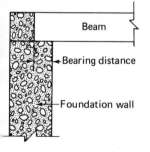

Fig. 3-7   Bearing Distance

Bearing Area

The beam must have sufficient bearing area on the foundation wall to avoid crushing the foundation wall or the wood beam under the concentrated load transferred to the foundation wall. A general rule states that the beam should have a minimum bearing distance of 5″ (see Fig. 3-7). A 5″ bearing distance is sufficient in most types of residential construction, but if an unusual condition exists and a large concentrated load is transferred to the wall the bearing area should be checked by applying standard engineering formulas.

To set the beam at the proper height or grade the length of the columns is adjusted. To straighten the beam from side to side temporary braces are set and adjusted between the foundation wall and the beam.

COLUMNS

A column may be defined as a post or vertical member designed to carry an important load. Columns which support the beam in residential construction carry the beam loads to the column footing. The spacing of the columns establishes the actual beam span.

Column Spacing

Maximum spacing of columns in residential construction should be kept down to 8′ or 10′ in two story buildings to avoid having large concentrated loads on a column footing and also to avoid the need for large, heavy beams which restrict the headroom in the basement. In single-story structures a spacing of 12′ is usually acceptable.

To illustrate the situation, if a beam supported joist spans of 16′ from each side and was supported on posts 15′ apart, each post would carry about 17 tons and would require larger footings than is common for many types of soil.

A 12″ I beam weighing 31.8 lb/ft would be needed for a 15′ span. A 45′ long beam of this size would weigh around 1400 lb and would be difficult to handle. If the beam span is cut to 8′ the column load is cut to about 9 tons and a 7 × 15.3 lb I beam weighing about 690 lb can be used. This beam would be easier to handle and would require smaller columns and column footings.

Columns supporting simple beams (see Fig. 3-8) support the beam load to the midpoint of the beam plus one-half the weight of the beam. If the load per beam span in Fig. 3-8 is 10,000 lb and the beam weighs 200 lb per span, the load on the

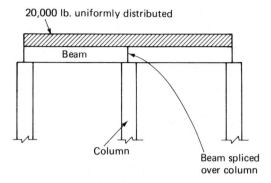

20,000 lb. uniformly distributed

Beam

Column

Beam spliced
over column

Beam wt. — 200 lb. per span
Column load — 10,200 lbs.

Fig. 3-8   Column Supporting Simple Beams

column would be

$$\frac{10{,}200}{2} + \frac{10{,}200}{2} \text{ or } 10{,}200 \text{ lb}$$

If the beam is continuous over the column as in Fig. 3–9, the column supports a greater proportion of the load. For all practical purposes, when a single column supports a continuous beam it carries the load to five-eighths of the span on either side of the column. Using the same loading as in Fig. 3–8 the column load would be

$$\tfrac{5}{8}\,(10{,}200) + \tfrac{5}{8}\,(10{,}200) \quad \text{or} \quad 12{,}750 \text{ lb}$$

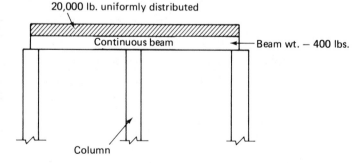

20,000 lb. uniformly distributed

Continuous beam

Beam wt. — 400 lbs.

Column

Column load — 12,750 lbs.

Fig. 3-9   Column Supporting Continuous Beams

When two or more columns are used to support a continuous beam the task of determining the column load becomes more complicated. In Fig. 3–10 a continuous beam is supported by two columns and a foundation wall at each end. Column A will carry one-half of the load in the center span and approximately five-eighths of the load in the span between the wall and the column, and column B will carry the other one-half of the load in the center span and approximately five-eighths of the load between the column and the foundation wall on the right.

In Fig. 3–11 three columns are used to support a continuous beam. Columns A and C carry approximately five-eighths of the load in the spans between the wall and the columns and one-half of the load between the columns. Column B carries one-half of the load between column A and column B and one-half of the load between column B and column C.

The continuous beam in Fig. 3–12 is supported by four columns and the foundation walls. Columns A and D carry approximately five-eighths of the beam load between the columns and the wall and one-half of the load in the spans between the columns. Columns B and C carry one-half of the beam load on either side of the columns.

When columns support continuous beams the five-eighths factor is not always exactly correct, but it is close enough to use for all practical purposes. Therefore, any time a continuous

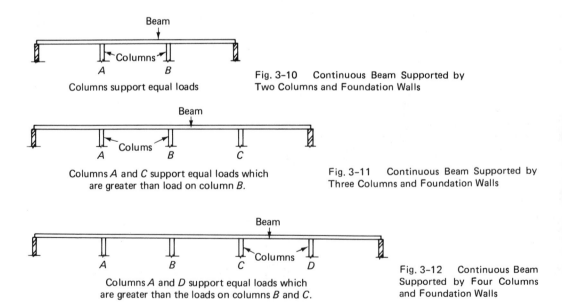

Beam

A B

Columns support equal loads

Fig. 3–10   Continuous Beam Supported by Two Columns and Foundation Walls

Beam

A B C

Columns A and C support equal loads which are greater than load on column B.

Fig. 3–11   Continuous Beam Supported by Three Columns and Foundation Walls

Beam

A B C D

Columns A and D support equal loads which are greater than the loads on columns B and C.

Fig. 3–12   Continuous Beam Supported by Four Columns and Foundation Walls

Table 3-3. Maximum allowable column loads

For Douglas fir, southern pine, and North Carolina pine, No. 1
common grade

Nominal size, inches	3 by 4	4 by 4	4 by 6	6 by 6	6 by 8	8 by 8
Actual size, inches	$2\frac{5}{8}$ by $3\frac{5}{8}$	$3\frac{5}{8}$ by $3\frac{5}{8}$	$3\frac{5}{8}$ by $5\frac{5}{8}$	$5\frac{1}{2}$ by $5\frac{1}{2}$	$5\frac{1}{2}$ by $7\frac{1}{2}$	$7\frac{1}{2}$ by $7\frac{1}{2}$
Area in square inches	9.51	13.14	20.39	30.25	41.25	56.25
Height of column:						
4′	8,720	12,920	19,850	30,250	41,250	56,250
5′	7,430	12,400	19,200	30,050	41,000	56,250
6′	5,630	11,600	17,950	29,500	40,260	56,250
6′6″	4,750	10,880	16,850	29,300	39,950	56,000
7′	4,130	10,040	15,550	29,000	39,600	55,650
7′6″		9,300	14,400	28,800	39,000	55,300
8′		8,350	12,950	28,150	38,300	55,000
9′		6,500	10,100	26,850	36,600	54,340
10′				24,670	33,600	53,400
11′				22,280	30,380	52,100
12′				19,630	26,800	50,400
13′				16,920	23,070	47,850
14′				14,360	19,580	44,700

Steel pipe columns—standard pipe safe loads in thousands of pounds

Nominal size, inches	6	5	$4\frac{1}{2}$	4	$3\frac{1}{2}$	3
External diameter, inches	6.625	5.563	5.000	4.500	4.000	3.500
Thickness, inches	0.280	0.258	0.247	0.237	0.226	0.216
Effective length:						
5′	72.5	55.9	48.0	41.2	34.8	29.0
6′	72.5	55.9	48.0	41.2	34.8	28.6
7′	72.5	55.9	48.0	41.2	34.1	26.3
8′	72.5	55.9	48.0	40.1	31.7	24.0
9′	72.5	55.9	46.4	37.6	29.3	21.7
10′	72.5	54.2	43.8	35.1	26.9	19.4
11′	72.5	51.5	41.2	32.6	24.5	17.1
12′	70.2	48.7	38.5	30.0	22.1	15.2
13′	67.3	46.0	35.9	27.5	19.7	14.0
14′	64.3	43.2	33.3	25.0	18.0	12.9
Area in square inches	5.58	4.30	3.69	3.17	2.68	2.23
Weight per pound per foot	18.97	14.62	12.54	10.79	9.11	7.58

beam is supported by two or more columns, the load on the
outer columns will be equal to the sum of five-eighths of the
combined beam load and beam weight for the outer span plus
one-half of the combined beam load and beam weight for the
inner span.

If columns are equally spaced it is only necessary to find

the load on the column nearest the foundation wall, since this column carries the greatest load. The size of column needed to support this load is then determined and is used for all the columns.

Before the size of a column can be determined, it is necessary to know the load on the column and the effective length of the column. Short columns will fail in compression while longer columns, when subjected to compressive forces, will also bend. Therefore a short column of a given size will support a greater load than a longer column of the same size material. Longer columns must be made of larger material to resist bending.

The narrowest dimension of a column generally determines its stiffness. The ratio of thickness to length for wood is usually 50. Therefore the safe height for a 2 by 4 or 2 by 6 is somewhere between 6' and 7'. Because of the tendency to bend in the narrow direction it is desirable to use wood columns which are square or round to obtain equal bending strength in two directions.

Wood columns may be solid timbers or they may be built up of two or more 2"-thick members. When they are built up care must be taken to fasten the pieces together so that they will work together as a unit. If they are not securely fastened they could fail individually. Nails or bolts may be used to fasten built-up column members together.

When the type of material, load, and effective length are known, the column size can be determined by using the proper table of column sizes. Table 3–3 gives column sizes for loads encountered in residential construction.

COLUMN INSTALLATION

Steel pipe columns are the most commonly used types of columns for residential construction. They may be hollow or filled with concrete and have bearing plates welded to each end. These columns are fastened to the beam by bolting through pre-drilled holes or by bending steel straps, which are welded to the top column-bearing plate, up around the flange of the beam (see Fig. 3–13).

Steel pipe columns are ordered to within 1" of their exact length. Steel shims are placed between the column base and the concrete footing to raise the beam to the proper elevation.

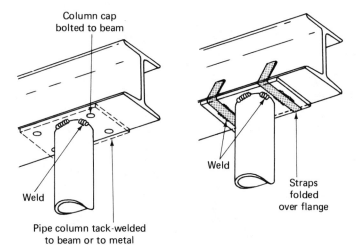

Column cap
bolted to beam

Weld

Pipe column tack-welded
to beam or to metal
cap-plate

Weld

Straps
folded
over flange

Fig. 3-13   Fastening Steel Column to Beam

These shims will vary in thickness from $\frac{1}{16}''$ to $\frac{1}{2}''$ and are usually 6'' square. The number of shims and thickness used depends on the distance the beam must be raised to bring it up to proper grade (see Fig. 3-14).

In some cases the lower end of the column is fitted with a large nut which uses a jack screw 1'' in diameter (see Fig. 3-15). When the jack screw is used a steel bearing plate with a center mark is placed on the footing and the jack screw is placed on the center mark. By adjusting the length of the screw the beam supported by the column may be raised to the proper elevation.

The center mark on the bearing plate serves to keep the jack screw from sliding during the adjusting operation. Advantages in using the jack screw column are that the need for shims is eliminated, adjustment of length is easy, and adjustment is infinite.

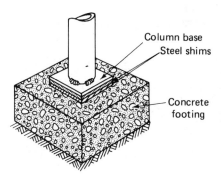

Column base
Steel shims

Concrete
footing

Fig. 3-14   Column Set on Steel Shims

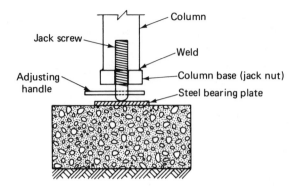

Fig. 3-15   Column Base with Jack Screw

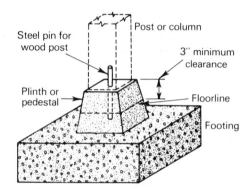

Fig. 3-16   Base of Wood Column

To keep the column from moving on the footing the concrete floor completely surrounds the column base and the jack screw. Where no concrete is poured around the columns it is necessary to provide anchor bolts in the footing and holes in the column base plate to anchor the column to the footing.

Wood columns are set on a concrete plinth block so that the bottom of the column is at least 3" above the basement floor. This is done to keep the column dry and to minimize the possibility of decay. A dowel pin is placed in the block and column to prevent the column being forced off the block (see Fig. 3-16).

Before installing, wood columns are cut to exact length. Therefore no shimming is necessary to bring the beam up to proper elevation. Care must be taken to cut the column square in both directions so that it will have full bearing area. If it is not cut square in both directions there will be crushing of the wood fibers at the high points, which will result of settling of the beam.

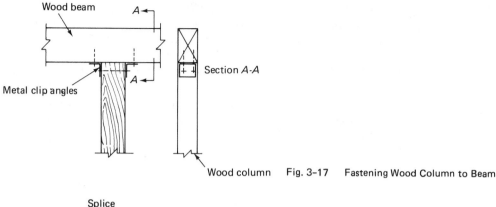

Fig. 3-17   Fastening Wood Column to Beam

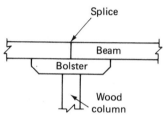

Fig. 3-18   Bolster Supporting Spliced Beam

The upper end of wood columns may be fastened to the beam by means of bolts, lag screws, nails, or metal angles (see Fig. 3-17). When a beam is spliced over the column a member known as a bolster is placed atop the column to provide sufficient bearing area for the beam to prevent crushing of the wood fibers (see Fig. 3-18).

FLOOR FRAMING

Floor joists are those framing members which actually carry the floor loads between supports. These joists are actually a type of beam which carries the floor load to the foundation walls and the central beam or girder.

Joist Spacing

The common spacings for floor joists are 12″, 16″, and 24″ on center. This means that when measuring from side to side or center to center the joists will be uniformly spaced to support the flooring. The spacings of 12″, 16″, and 24″ on center make it possible to use 4′ by 8′ sheets of plywood for subflooring and have all end joints fall on the center of a joist without the need for cutting the sheets to length. Joist spacings of 13.7″ and 19.2″ on center can also be used. These spacings will place a joist at 8′ multiples. They are used to exploit the maximum strength of a joist and permit fewer joists to carry the floor load (see Fig. 3-19). This spacing also allows for use of 96″-long plywood sheets.

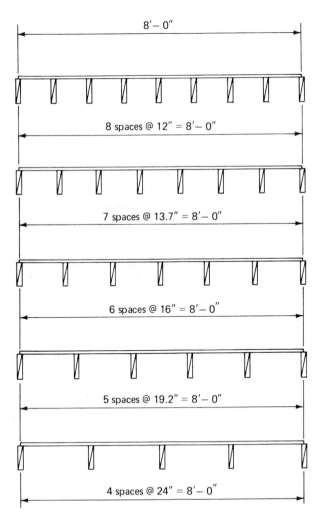

Fig. 3-19   Comparative Joist Spacings

Joist Characteristics

Floor joists must have two characteristics: strength and stiffness. A joist is said to have strength if it safely supports the load placed on it. If it sags or deflects unduly while supporting a load it lacks stiffness. Sufficient stiffness is necessary to avoid cracking and nail popping in the finished walls and ceilings supported by the joists. It is also necessary to avoid undue springiness or bouncing of the floor when people or objects move across the room.

Floor supported by joists which lack sufficient stiffness give people an uneasy feeling when they walk across the room, and movement in the floor often causes lamp tables to move and objects on the tables to fall.

Residential floors are usually designed to carry a live load of 40 lb/sq. ft. Most building codes specify this requirement and have tables of joist sizes which are used to determine joist size for various spans.

The span of joist is the unsupported distance from the inside face of the foundation wall to the center of the beam at the inner end. Joist size may be determined from the table when span, spacing, and live load per square foot are known. It is also necessary to know if the floor joists support a finished ceiling on the underside in order to choose the proper table (see Tables 3-4 and 3-5).

Types of Sills

Two types of sills are commonly used in platform framing. They are known as the L-type box sill and the "plain sill" (see Fig. 3-20). When the L-type sill is used a 2 by 6 or a 2 by 8 is placed on top of the foundation wall. This sill may be set in a bed of cement mortar and bolted to the foundation wall, or it may be placed on a layer of insulation which seals the space between the wood and the foundation wall. In some cases no insulation is used.

In areas where winds do not require the anchoring of the floor frame to the foundation, the sill is not bolted to the foundation wall, and it is possible that the sill plate may be left out entirely. This type of construction may be referred to as a plain box sill.

Table 3-4. Maximum spans for joists (uniformly loaded). Extreme fiber stress in bending 1200 lb/ sq. in (no plastered ceiling below)

Live load—pounds per square foot	Spacing	2" wide by depth of—					3" wide by depth of—				
		6	8	10	12	14	6	8	10	12	14
30	12	13- 5	17- 8	22- 2	24- 0	—	16- 8	21-10	24- 0	—	—
	16	11- 9	15- 6	19- 5	23- 3	24- 0	14- 7	19- 3	24- 0	—	—
	24	9- 8	12-10	16- 2	19- 5	22- 6	12- 1	16- 0	20- 1	24- 0	—
40	12	12- 0	15-11	19-11	23-11	—	15- 0	19- 8	24- 0	—	—
	16	10- 6	13-11	17- 4	20-11	21- 0	13- 1	17- 4	21- 8	24- 0	—
	24	8- 7	11- 5	14- 5	17- 5	20- 3	10-10	14- 4	18- 0	21- 8	24- 0
50	12	10-11	14- 5	18- 2	21-11	24- 0	13- 8	18- 0	22- 6	24- 0	—
	16	9- 6	12- 7	15-10	19- 1	22- 3	11-11	15- 9	19- 9	23- 9	24- C
	24	7-10	10- 4	13- 1	15- 9	18- 5	9- 9	13- 0	16- 5	19- 9	23- 0
60	12	10- 0	13- 4	16-10	20- 2	23- 6	12- 7	16- 8	20-11	24- 0	—
	16	8- 9	11- 8	14- 8	17- 8	20- 7	11- 0	14- 6	18- 4	22- 0	24- 0
	24	7- 3	9- 6	12- 1	14- 7	17- 0	9- 0	12- 0	15- 2	18- 3	21- 4
70	12	9- 5	12- 5	15- 8	18-11	22- 0	11-10	15- 7	19- 7	23- 6	24- 0
	16	8- 1	10-10	13- 8	16- 6	19- 3	10- 4	13- 7	17- 1	20- 7	24- 0
	24	6- 8	8-11	11- 3	13- 7	15-11	8- 5	11- 3	14- 1	17- 1	19-11
80	12	8- 9	11- 8	14- 9	17- 9	20-10	11- 1	14- 9	18- 6	22- 3	24- 0
	16	7- 8	10- 2	12-10	15- 6	18- 3	9- 9	12-10	16- 3	19- 6	22- 9
	24	6- 5	8- 5	10- 7	12- 9	15- 0	8- 0	10- 7	13- 4	16- 1	18-10

Table 3-5. Maximum spans for joists (uniformly loaded). Fiber stress 1200 lb/sq. in; modulus of elasticity 1,600,000 [plastered ceiling below (deflection not over 1/360 of span)]

Live load—pounds per square foot	Spacing	2" wide by depth of—					3" wide by depth of—				
		6	8	10	12	14	6	8	10	12	14
10	12	12- 9	16- 9	21- 1	24- 0	—	14- 7	19- 3	24- 0	—	—
	16	11- 8	15- 4	19- 4	23- 4	24- 0	13- 6	17- 9	22- 2	24- 0	—
	24	10- 3	14- 6	17- 3	20- 7	24- 0	11-11	15- 9	19-10	23- 9	24- 0
20	12	11- 6	15- 3	19- 2	23- 0	24- 0	13- 3	17- 6	21- 9	24- 0	—
	16	10- 5	13-11	17- 6	21- 1	24- 0	12- 0	16- 1	20- 2	24- 0	—
	24	9- 2	12- 3	15- 6	18- 7	21- 9	10- 6	14- 2	17-10	21- 6	24- 0
30	12	10- 8	14- 0	17- 9	21- 4	24- 9	12- 4	16- 4	20- 5	24- 5	—
	16	9- 9	12-11	16- 3	19- 6	22- 9	11- 4	14-11	18- 9	22- 7	26- 4
	24	8- 6	11- 4	14- 4	17- 3	20- 2	10- 0	13- 2	16- 8	19-11	23- 4
40	12	10- 0	13- 3	16- 8	20- 1	23- 5	11- 8	15- 4	19- 3	23- 1	26-11
	16	9- 1	12- 1	15- 3	18- 5	21- 5	10- 8	14- 0	17- 8	21- 3	24-10
	24	7-10	10- 4	13- 1	15- 9	18- 5	9- 4	12- 4	15- 7	18- 9	22- 1
50	12	9- 6	12- 7	15-10	19- 1	22- 4	11- 0	14- 7	18- 4	22- 0	25- 8
	16	8- 7	11- 6	14- 7	17- 6	20- 5	10- 0	13- 4	16-10	20- 3	23- 8
	24	7- 3	9- 6	12- 1	14- 7	17- 0	8-10	11- 9	14-10	17-10	20-10
60	12	9- 0	12- 0	15- 2	18- 3	21- 4	10- 6	14- 0	17- 7	21- 1	24- 7
	16	8- 1	10-10	13- 8	16- 6	19- 3	9- 7	12-10	16- 1	19- 4	22- 7
	24	6- 8	8-11	11- 3	13- 7	15-11	8- 5	11- 3	14- 1	17- 0	20- 0
70	12	8- 7	11- 6	14- 6	17- 6	20- 6	10- 1	13- 5	16-11	20- 5	23- 9
	16	7- 8	10- 2	12-10	15- 6	18- 3	9- 3	12- 3	15- 5	18- 7	21-10
	24	6- 5	8- 5	10- 7	12- 9	15- 0	8- 0	10- 7	13- 4	16- 1	18-10

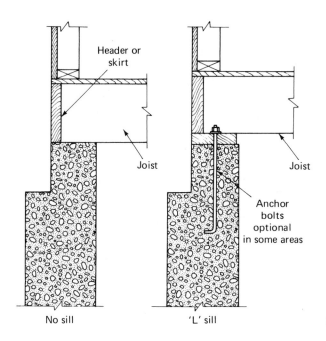

No sill   'L' sill   Fig. 3-20   Box Sills for Platform Framing

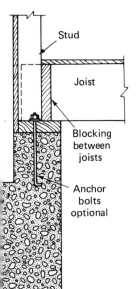

.3-21   T-Type Box Sill
Balloon Framing

In either the plain box sill or the L-type box sill without anchor bolts, the weight of the building is depended on to keep it on the foundation. Where high winds are no problem this is a completely satisfactory method of construction.

Balloon framing requires a T-type box sill. This sill gets its name from the inverted T formed by the sill plate and the blocking between the joists (see Fig. 3-21). As with platform framing, the sill may be set in a bed of mortar, laid on a layer of insulation, or placed directly on the foundation wall. Anchor bolts may or may not be used, depending on local practice.

Sill Layout

Before beginning sill layout the building foundation dimensions should be checked and foundation diagonals measured to determine building squareness. After checking the dimensions, the location of the inside edge of the sill should be marked on top of the foundation wall at each end of the building, and a chalk line should be snapped between the marks. This line serves as a base line from which to measure in locating the anchor bolt holes.

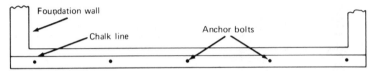

Step 1   Snap chalk line on top of foundation
         to show location of inside edge of sill

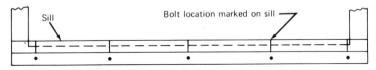

Step 2   Place sill on top of wall and
         mark bolt locations

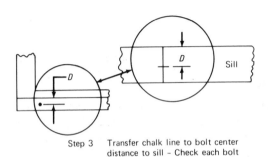

Step 3   Transfer chalk line to bolt center
         distance to sill – Check each bolt
         for variation

Fig. 3-22   Sill Layout

The sill plate should be cut to length, placed on top of the foundation wall against the anchor bolts, and the end located according to previously established layout marks. With the aid of a square the location of both sides of the bolts should be marked on the sill, and the distance from the chalk line to the center of each bolt measured and transferred to the sill (see Fig. 3-22). Following the layout, holes are drilled, and the sill is put in place. If layout has been carefully done the inside edge of the sill will coincide with the chalk line.

Sill Sealer

To seal the joint between the wood framework and the masonry, fiberglass insulation may be placed on top of the foundation wall before the wood sill and joists are installed

Fig. 3-23   Sill Sealer (Courtesy Owens-Corning Fiberglas Corporation)

(see Fig. 3–23). This sealer is inexpensive and fills in the irregularities between the wood framing and the top of the foundation wall. The sill seal provides a barrier to air infiltration, and thereby reduces drafts and cuts heating costs.

The sill sealer is easily installed, but care should be taken to roll it out uniformly, avoiding stretching or bunching. When the wood sill is placed over it, the sealer should span the full width of the sill without gaps. When termite shields are used, the sill sealer is placed between the termite shield and the wood sill.

Termite Shields

In areas where termites are a problem, termite shields or chemical treatment of the soil and framing lumber may be used to prevent infestation and damage.

When metal shields are used the top of the foundation should be at least 8″ above the outside finish grade. These shields should be bent down at a 45° angle and project at least 2″ beyond the foundation wall. All joints or seams should be lock folded or completely soldered to prevent the passage of termites. Where bolts pass through the shield a heavy layer of coal-tar pitch should be applied around the bolt to seal the area completely.

The termite shield in Fig. 3–24 would be used on a frame

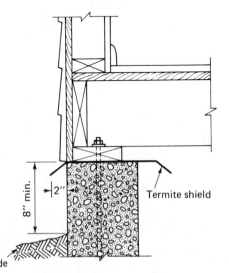

8″ min.

2″

Termite shield

Fin. grade

Fig. 3-24 Termite Shield for Frame Wall

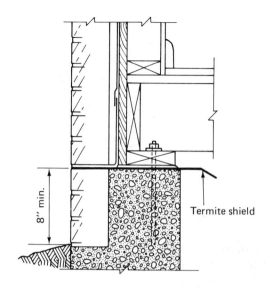

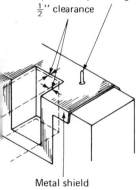

Fig. 3-25   Termite Shield Installed in Masonry Veneer

Penetrations for anchor-bolts etc., shall be sealed with solder, coal tar pitch or by welding

$\frac{1}{2}''$ clearance

Metal shield

Fig. 3-26   Termite Shield for Wood Beam Pocket

wall. Care must be taken to avoid bending the shield down against the wall on the outside, because this would render the shield ineffective.

Walls with masonry veneer may have termite shields applied as shown in Fig. 3-25. When the shield is imbedded in a mortar joint a mortar rich in cement should be used, because termites may bore through weak mortar.

The ends of wood beams should be protected by a metal shield as shown in Fig. 3-26. This shield should have folded or soldered joints and should be connected to the continuous shield by folded or continuously soldered joints.

Where chemicals are used to treat soil surrounding the building against termite infestation they should be applied by contractors having experience in their application. These chemicals should be applied in strict accordance with the manufacturer's recommendations and local regulations.

Framing lumber may be treated with coal-tar creosote, pentachlorophenal, copper naphthenate, and other materials to make them termite and decay resistant. The preservative may be applied to the lumber by spraying, brushing, or soaking at the job site, or it may be applied under pressure at a wood preserving plant.

Lumber treated at a wood preserving plant can be expected to be more resistant to decay and termites because of the greater quantity of preservative which has been absorbed into the wood.

Bearing Distance

Joists should have a minimum end bearing distance of 3″ on concrete or concrete block. A 4″ end bearing distance is preferred and should be used when possible. A 2″ end bearing distance is usually acceptable when the joists bear on wood or on steel. However, because there is some tendency for the wood fibers in the bearing area to become crushed a 3″ to 4″ bearing area is preferred.

Framing Joists in Masonry Walls

Joists framing into masonry walls must have a "fire cut" (see Fig. 3-27). This fire cut is made so that the joist has sufficient bearing area at the lower edge and is angled up so that the top edge of the joist is not enclosed in the masonry wall. The purpose of this cut is to allow the joists to be released from the wall in the event they are burned through during a fire.

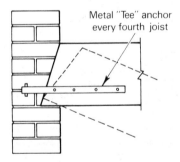

Metal "Tee" anchor every fourth joist

Fig. 3-27   Firecut on Joist Framed into Masonry Wall

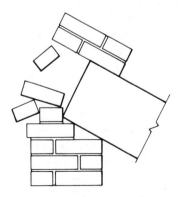

Fig. 3-28   Joist Improperly Framed into Masonry Wall

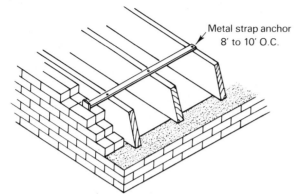

Fig. 3-29   Anchoring Masonry Walls Parallel to Floor Joists

Without a fire cut the end of the joist would cause the wall to collapse in the event the joist was burned through and forced down during a fire (see Fig. 3-28).

When the floor frame is 5′ or more above grade it is necessary to install strap anchors at the bottom edge of every fourth joist for the purpose of tying the opposite walls together to prevent them from moving away from the floor. On side walls strap anchors are nailed to the top of the floor joists at 8′ to 10′ intervals (see Fig. 3-29).

Framing Joists at Beam

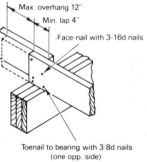

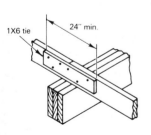

Fig. 3-30   Framing Joists to Top of Wood Beam

The best way to frame joists to the beam is simply to rest them on top of the beam. This provides full support for the joist and easy framing, since no special cutting is necessary. In Fig. 3-30 the joists are framed to the top of wood beams. This makes it possible to nail the joists to the beam to keep them aligned before the subflooring is placed. Notice that when the joists are lapped the minimum lap is 4″ and the maximum overhang is 12″. An overhang of more than 12″ will cause the flooring to be loosened, because deflection at the center of the joist span will cause the overhang to move upward. The constant movement will cause the floor over the overhang to become loose and squeak (see Fig. 3-31). It is good practice to limit the overhang to 2″ when the joists are lapped.

When the joists are set in line the ends of the joist should fall at the center of the beam, and the joists should be tied together with a 1 by 6, a metal strap, or a patented fastener.

The joists in Fig. 3-32 are lapped over a steel beam. The

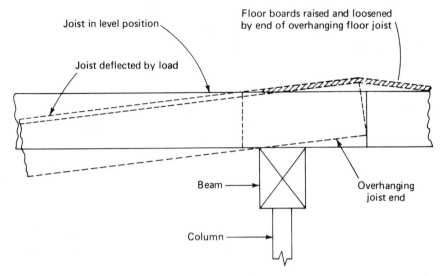

Fig. 3-31   Joist Overhand

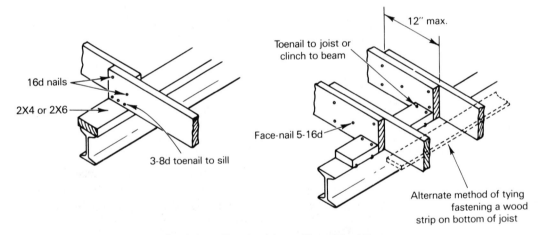

Fig. 3-32   Framing Joists to Top of Steel Beam

wood sill on the beam provides a nailing surface and also helps to keep settling due to shrinkage of framing lumber equal when used in conjunction with a sill on the foundation walls.

When joists are framed directly to the steel some building codes require that spacer blocks be placed between the joists or that a tie be nailed to the bottom of the joists.

In some cases placing the beam below the joists may be objectionable because it reduces the headroom in the basement.

To minimize or do away with this loss of headroom the joists may be framed to the side of the beam. In the case of a wood beam the joists may rest on a ledger or be supported by metal stirrups (see Fig. 3-45).

Joists framed to the side of a steel beam may rest on the lower flange or on a bearing plate which has been welded on the lower flange. The joists on the opposite sides of the beam are tied together to keep them from sliding off the flange during the construction operation (see Fig. 3-33).

Joist Layout

The location of all floor joists is marked off on the header joist or skirt in platform framing and on the sill in balloon framing. The layout procedure is similar for both types of framing, so only the layout procedure for platform framing will be discussed here.

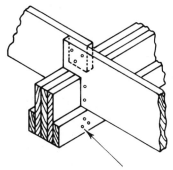

Ledger board

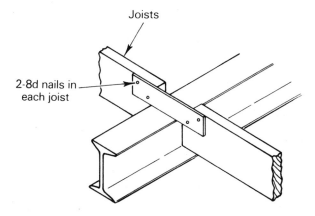

Joists

2-8d nails in each joist

Fig. 3-33   Framing Joists to Side of Beam

Straight material for header joists is selected from the lumber available on the job and placed on top of the foundation wall or at some other convenient location. If the joists are 16″ on center the carpenter locates a joist at the end of the header and the side of a second joist 15¼″ in from the end of the header. By starting his layout in this manner the carpenter will have a distance of 16″ from the outside of the end joist to the center of the second joist. The remaining regular joists are marked off 16″ from side to side, which makes them 16″ on center and places the center of the seventh joist 96″ or 8′ from the end of the floor frame to accommodate the end of an 8′ plywood sheet (see Fig. 3–34).

When the joists are 12″ on center the side of the second joist is placed 11¼″ from the end of the header, and when they are 24″ on center the second joist is placed 23¼″ from the end.

Joists may be located by using the carpenter's framing square and pencil for measuring and marking, or a steel tape and a combination square and pencil. The framing square layout is satisfactory for residential work if care is exercised in marking. (Otherwise there will be a tendency to gain ¹⁄₁₆″ or more on every step; see Fig. 3–35.) The use of the steel tape removes this error and is desirable for use on long buildings (see Fig. 3–36).

After the regular joists have been located and the header joist is cut to length it is necessary to lay out the locations of all partitions which run parallel to the joists so that extra joists may be installed to carry the weight of these partitions.

The extra joists should be installed as near the partition as possible. They may be nailed directly to the joist nearest the partition, or nailed to spacer blocks which have been fastened to the nearest regular joist. In some cases the extra joist is placed so that it stands alone (see Fig. 3–37). Whichever method is used, care should be taken to avoid placing the joist directly

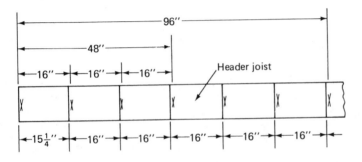

Fig. 3-34   Joist Layout

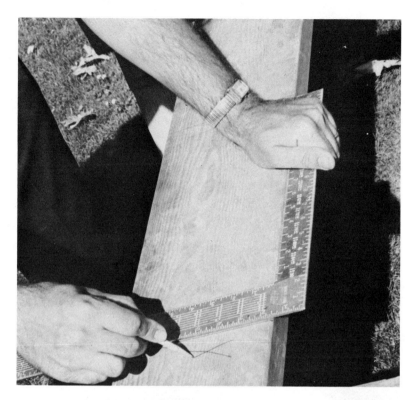

Fig. 3-35 Pencil and Square Layout

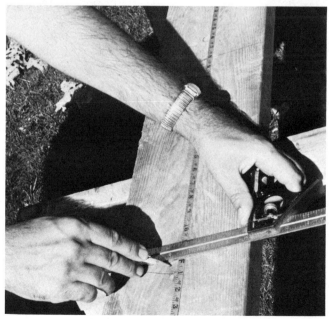

Fig. 3-36 Steel Tape and Combination Square Layout

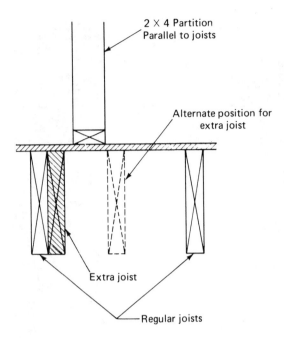

2 X 4 Partition
Parallel to joists

Alternate position for
extra joist

Extra joist

Regular joists

Fig. 3-37   Joist Supporting Partitions

under the partition. This is especially important when the partition will carry piping for plumbing and heating systems.

In lieu of extra joists, some building codes permit the use of solid bridging placed on 4′ centers in the joist space below partitions parallel to the joists. Layout must also be made for stairwell location and framing, chimney or fireplace framing, and the location of exterior doors (see Fig. 3-38).

Joist Installation

After the layout is completed the joists are set in place on the foundation wall and nailed to the header joist in accordance with the layout marks. Before the joists are nailed to the header they are checked for camber or crown. Because the joists will have a tendency to deflect over a period of time, the crown is placed upward. As deflection takes place the tops of the joists, which establish the level of the floor, will become nearly

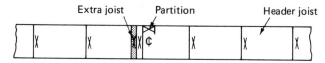

Layout for partition parallel to joists

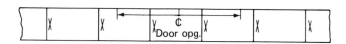

Layout for stairwell

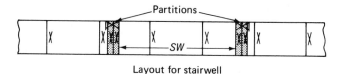

Firplace layout

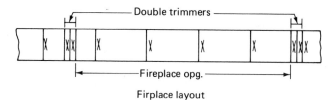

Door layout   Fig. 3-38   Joist Layouts

straight. It is important that the top edge of all the joists be kept flush (even) with the top edge of the header joists so that the floor will be reasonably straight.

The joists are nailed to the header with three to five 16d nails. Various building codes have different nailing requirements. Some codes require that 20d nails be used. Regardless of the size of nail, a sufficient number should be used to attain required strength.

In balloon framing the joists are toenailed to the sill and blocking is nailed between the joists to act as a firestop and to provide support for the ends of the floor boards. This blocking is nailed along the inside wall line with sufficient space allowed for the studs to be nailed to the sill between the edge of the sill and the blocking.

Second-Floor Joists

In platform framing the second-floor joists are toenailed to the top plate of the first-floor walls, and a header joist is nailed over the ends of the joists (see Fig. 3–39).

In balloon framing the joists are set on a ribbon and nailed to the side of the studs. Three to five 16d nails are used to fasten the joists to the studs (see Fig. 3–40).

Ceiling Joists

Floor joists which carry a ceiling on the underside serve a double purpose and are designed to carry the floor load and the weight of the ceiling material. Joists which carry only the weight of the ceiling are known as ceiling joists. They must be stiff enough to resist bending, which would cause the plaster or other ceiling material to crack.

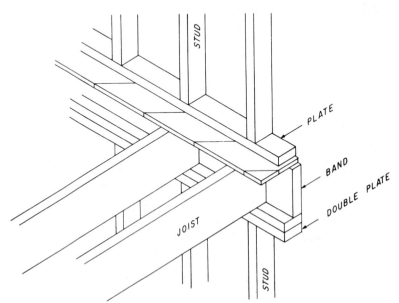

Fig. 3–39 Framing Second Floor Joists to Outside Wall in Platform Framing (Courtesy National Forest Products Association)

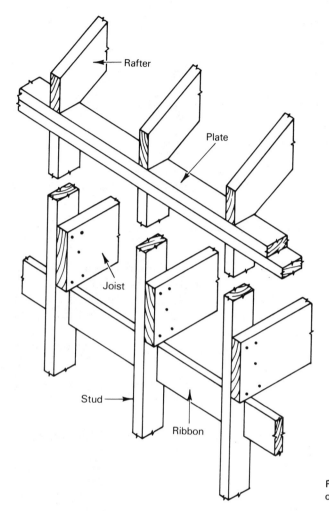

Fig. 3-40   Second Floor Joists Supported on Ribbon in Balloon Framing

FRAMING FLOOR OPENINGS

Openings in the floor frame for the stairways, chimneys, and fireplaces require the cutting of some of the regular joists to make the proper-sized opening. The portion of the regular joist remaining in the floor frame is known as a tail joist and is supported by a header joist. The header hoist is supported by a trimmer joist (see Fig. 3-41).

The tail joist will be the same size as the regular joists. That is, if 2 by 10's are used for the regular joists, following

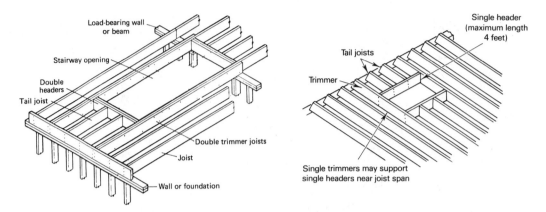

Fig. 3-41   Typical Floor Openings

common practice, 2 by 10's would be used for the tail joists. The depth of the header and trimmer joists is also regulated by the depth of the regular joists. Where added strength is required the width of headers and trimmers is increased. Depth usually cannot be increased because it would result in a projection below the ceiling.

Header Joists

The header joist supports the floor load to the midpoint of the tail joists it carries. This floor load on the header includes the live load (usually 40 lb/sq. ft) and the weight of the floor construction (usually 10 lb/sq. ft), and for all practical purposes the load on the header joist may be considered as being uniformly distributed. Therefore the load on the header joist may be determined in a manner similar to that for calculating beam load.

The header in Fig. 3-42 supports tail joists which are 15' long. Of the entire load on the tail joists, one-half is supported by the beam and one-half is supported by the header. The distance from the header to the midpoint of the tail joists may be referred to as the header load width.

If the total combined live load and dead load is 50 lb/sq. ft the load per lineal foot on the header is 50 × 7½ or 375 lb. The total load on the header can be found by multiplying the load per lineal foot by the length of the header. For the header in Fig. 3-42 it is 8 × 375 or 3000 lb. When the joist header span

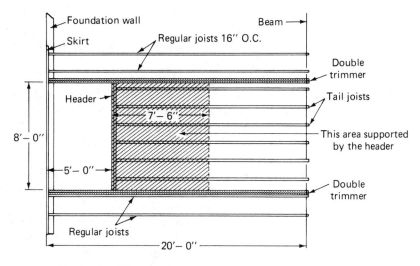

Live load — 40⁼ per sq. ft.
Load carried by header — 50⁼ per sq. ft.
Total header load — 50 x 7½ x 8 = 3,000 lbs.
Trimmer load — 1,500 lbs. concentrated at ¼ span

Fig. 3-42   Determining Header and Trimmer Load

and total load are known, the size can be determined by referring to tables of wood beam and girder sizes.

The procedure for finding the load and size of a header
joist may be summarized as follows:

1. Determine load per square foot carried to header.

2. Determine header load width:

$$\frac{\text{Length of tail joist}}{2}$$

3. Determine load per lineal foot of header.

4. Find length of header—header span.

5. Find total load on header.

6. Find header size in table.

Trimmer Joists

Trimmer joists carry a concentrated load from the header
joist. Of the entire load on a header joist, one-half is concen-

trated on the support at each end. The trimmer joist in Fig. 3–42 supports a concentrated load of 1500 lb. The load is supported 5′ from the end of a 20′ trimmer joist, and therefore it is concentrated at $^5/_{20}$ or ¼ span.

Concentrated loads of a given size have a different effect on beams than uniformly distributed loads of the same total size. Also, a concentrated load applied at one-half of the span or center of a trimmer acts differently than one applied at one-quarter of the span. Therefore the location of the concentrated load must always be known.

A concentrated load at the center of a span will cause bending twice as great as a uniformly distributed load of the same total size. As the concentrated load is moved to the side the amount of bending decreases proportionally, until it is zero directly over the support. The comparison of bending caused by a uniformly distributed and concentrated load applied at various points is shown in Fig. 3–43.

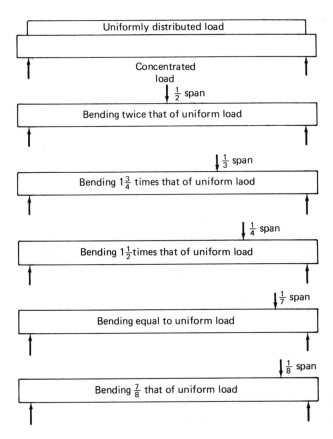

Fig. 3–43   Comparison of Bending Effect – Uniformly Distributed and Concentrated Loading

Beam tables are usually set up for uniformly distributed loads. Therefore it is necessary to find the size of a uniformly distributed load at a given point before the beam tables can be used to find the size of a beam or trimmer joist carrying a concentrated load.

The concentrated load of 1500 lb at one-quarter of the span in Fig. 3–42 would cause bending one and one-half times that of a uniformly distributed load (see Fig. 3–43). The equivalent uniform load would be one and one-half times 1500 or 2250 lb. This equivalent load would then be found in the table to determine the size of trimmer needed.

The procedure for finding the load and size of a trimmer joist may be summarized as follows:

1. Determine the concentrated load on the trimmer.
2. Determine the location of the concentrated load.
3. Determine the equivalent uniform load.
4. Find trimmer size in beam table.

Framing an Opening

The first step in framing an opening in the floor is to install single trimmer joists adjacent to the opening. Usually the location of the header joist will be marked on the trimmers before they are installed. Next a single header is nailed between the trimmers and fastened with four 16d or three 20d nails (see Fig. 3–44). After the header is nailed in place the tail joists are fastened to the header using four 16d nails.

When all the tail joists are in place the header may be doubled. The double header is nailed to the trimmer joist in the same manner as the single header and then fastened to the single header with 16d nails placed near the edge and placed alternately every 6″. This places the nails 12″ apart at each edge.

Double trimmers are installed after the double headers are in place and nailed together using 16d nails spaced every 6″ alternately near the top and bottom edges. The connection between the header and trimmer is subjected to concentrated loading. Nails alone may be insufficient to support the header load. Several types of joist hangers are available for use in providing extra support for headers and tail joists.

Perhaps one of the simplest is the U type, which may be nailed to the header and tail joist or to the header and trimmer

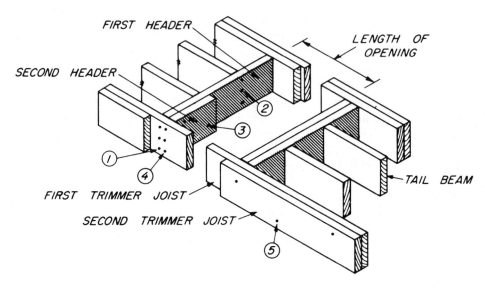

FIRST HEADER

SECOND HEADER

LENGTH OF OPENING

② ③ ① ④

FIRST TRIMMER JOIST

SECOND TRIMMER JOIST

TAIL BEAM

⑤

Fig. 3-44   Framing Floor Openings

(see Fig. 3–45). These hangers are made in several sizes and will adequately support any loads normally encountered in residential construction.

BRIDGING

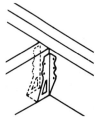

Bridging is used to stiffen the floor frame. Its main purpose is to transfer a concentrated load from point of application to joists near the load (see Fig. 3–46).

Bridging may be of the crisscross type made of 1 by 3 or 1 by 4 stock cut on the job, or it may be metal bridging of any one of a variety of types available on the market. The wood bridging transfers load through compression of the members. Some metal bridging transfers the load by tension of its members while other types transfer load by compression (see Fig. 3–47).

Most metal bridging is more economical to install than wood bridging, but wood bridging usually gives better performance and may be more desirable.

A third type of bridging is solid bridging made up of pieces of joist material cut to fit snugly between the joists. The bridging is sometimes used with crisscross bridging in joist spaces which are smaller or larger than the normal spacing. These odd-

Fig. 3-45   U-Grip Joist Hanger

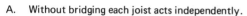

A.   Without bridging each joist acts independently.

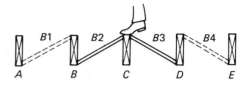

B.   Bridging helps spread load to adjacent joists.

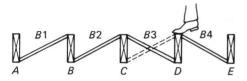

C.   Bridging must be criss-crossed to be efficient.

Fig. 3-46   Action of Bridging

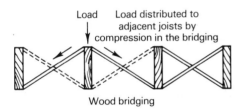

Load   Load distributed to
        adjacent joists by
        compression in the bridging

Wood bridging

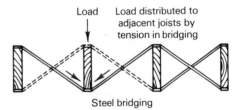

Load   Load distributed to
        adjacent joists by
        tension in bridging

Steel bridging

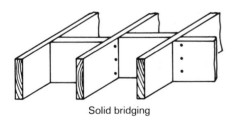

Solid bridging

Fig. 3-47   Tension and Compression in Bridging

sized spaces occur when joists are doubled, trimmers are installed, or joists are moved to make room for piping. Solid
bridging is also installed in places where it would be impossible
to nail the lower end of the bridging, such as spaces over the
foundation walls.

Bridging Layout

The length of wood bridging may be determined by holding the framing square on the edge of a piece of stock with the width of the joist on one side and the space between the joists on the other side (see Fig. 3–48). It is good practice to use a dimension ¼" less than the width of the joist when making the layout to allow for shrinkage of the joist material.

After the length and angles have been laid out a jig may be set up to cut the bridging using a portable power saw or a radial-arm saw.

Bridging Installation

As a general rule bridging is installed in joist spans greater than 8′. In spans between 8′ and 16′ one row of bridging is placed at the center of the span. When the span is over 16′ two rows of bridging are required. These rows of bridging are placed at one-third of the joist span.

When wood crisscross bridging is used, nails are started in

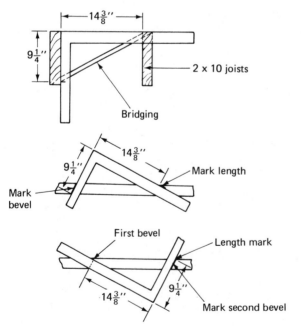

Fig. 3-48   Length of Bridging

each end before the bridging is installed. Starting the nails in this manner facilitates installation. The upper end of the bridging is nailed flush with the top of the joists on either side of a chalk line snapped on the joists. The lower end of the bridging is not nailed in place until after the subflooring is in place. The subflooring brings the top of the joists into alignment, and the deferred nailing allows the joists to lose some of their crown.

SUBFLOORING

The subfloor in residential construction is a wood floor which is laid over the floor joists and provides the base over which the finish floor is installed. There are two types of subflooring in general use. These are nominal 1″ boards in 6″ to 10″ widths and 4′ by 8′ sheets of plywood in various thicknesses.

Subflooring serves a number of purposes. It increases the strength of a floor and makes it possible to lay a fairly thin finish floor such as parquet or strip flooring. When board subflooring is laid diagonally it serves to brace the floor frame and stiffen the building. This is especially true on the upper floors. Laying the boards diagonally also allows the finish floor to be laid at right angles to the joists for maximum stiffness in the floor.

During the construction of the building the subfloor provides a working surface, thus saving the finish floor from possible damage. In the completed building the subfloor helps prevent dust from passing from floor to floor and also adds to the heat- and sound-insulating qualities of the floor.

Installing Board Subflooring

Boards should be installed with the end joints occurring at the center of a joist. This is required whether the boards are applied diagonally or at right angles to the joists.

To install boards diagonally the following procedure may be employed (see Fig. 3–49):

1. Make a mark 10′ from the corner of the building on the header joist and on the end joist.

2. Drive temporary nails at each mark.

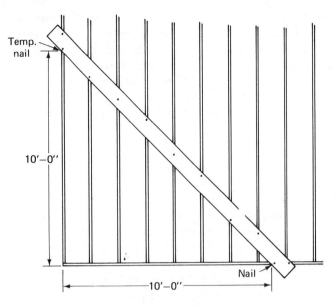

Temp.
nail

10'–0"

Nail

|←————10'–0"————→|

Start with straight 16' board or snap chalk
line between temporary nails and align board
with chalk line

Fig. 3–49   Starting Board Subflooring

3. Snap chalk line on top of joists between nails.

4. Select straight 16' floor boards.

5. Nail boards to joists and straighten to follow chalk line.

6. Install boards to corner leaving approximately ⅛" space between boards to allow for swelling in the event of rain before the roof is applied. (Allow ends to "run wild.")

7. Snap chalk line on boards in line with outside of header and end joist.

8. Trim wild ends.

9. Continue laying floor boards across the remaining area.

10. Trim ends periodically and use cutoffs as much as possible.

When the boards are being placed they are nailed lightly (tacked in place) with only enough nails to keep them from moving. After all the boards are in place the floor is nailed completely with two or three nails at each joist crossing. Automatic nailers may be used to facilitate this complete nailing.

Installing Plywood Subflooring

To assure alignment of the plywood sheets from one end of the building to the other a chalk line is snapped on the top edge of the joists 48″ in from the outside of the header joist (see Fig. 3-50). The plywood sheets are aligned with the chalk line and nailed to the joists. All end joints must be staggered and must occur at the center of a joist. After all the sheets have been tacked in place, they are nailed completely in accordance with local codes.

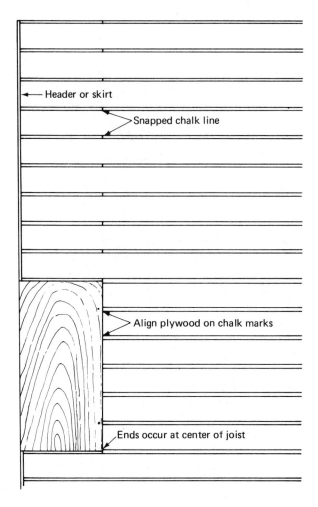

Header or skirt

Snapped chalk line

Align plywood on chalk marks

Ends occur at center of joist

Fig. 3-50   Starting Plywood Subfloor

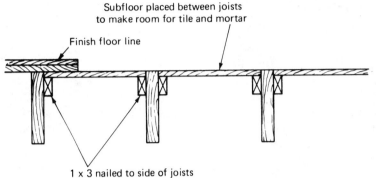

Subfloor placed between joists
to make room for tile and mortar

Finish floor line

1 x 3 nailed to side of joists

Fig. 3-51   Dropped Sub-
floor for Ceramic Tile or
Slate

Installing Dropped Subfloor

When ceramic tile or slate is set in cement mortar it is necessary to provide space for the thickness of the tile and mortar underlayment (see Fig. 3-51). This is done by placing the subflooring between the joists and supporting it on 1 by 3 or 1 by 4 material which is nailed to the joists with 8d nails.

NEW METHODS OF FLOOR FRAMING

Several new methods of building the floor frame have been developed to provide a floor which is sturdy and more economical than conventional floors. These new systems have had varying degrees of success, but only the test of time will determine their continued use.

One of these systems makes use of a single floor member which acts as a subfloor and underlayment and is placed over conventionally spaced joists (see Fig. 3-52). A second type of subfloor-underlayment construction uses joists spaced 32" or 48" on center with a plywood floor panel 1⅛" thick. The long edges of the plywood are tongue and groove. This material is called 2-4-1 (see Fig. 3-53).

Various stressed skin floor panels are in the experimental stages of development, and the U.S. Forest Products Laboratory has been conducting tests on "sandwich" panels and other types of floor construction.

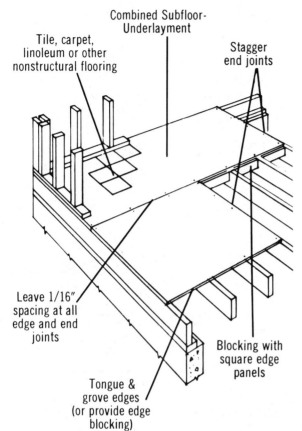

Combined Subfloor-
Underlayment

Tile, carpet,
linoleum or other
nonstructural flooring

Stagger
end joints

Leave 1/16"
spacing at all
edge and end
joints

Blocking with
square edge
panels

Tongue &
grove edges
(or provide edge
blocking)

Fig. 3-52 Combined Subfloor – Underlayment (Courtesy American Plywood Association)

APA Glued Floor System

The American Plywood Association glued floor system utilizes tongue and groove underlayment glued to the floor joists at the job site (see Fig. 3-54). The system uses fewer nails than conventional framing without glue and results in a stronger floor. This increased strength allows greater loads to be carried by smaller framing members. Gluing also eliminates squeaks in the floor that result from loose nails.

When using the glued floor system it is important that the framing lumber be clean and dry. The adhesive will not stick to muddy, dusty, or wet surfaces. Directions provided by the adhesive manufacturer should be followed carefully. Failure to follow these directions will result in an inferior job.

Fig. 3-53　2-4-1 Subfloor – Underlayment (Courtesy American Plywood Association)

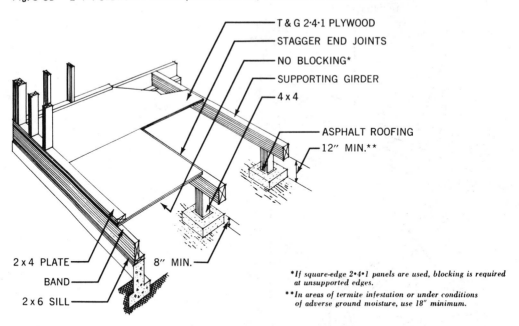

T & G 2·4·1 PLYWOOD

STAGGER END JOINTS

NO BLOCKING*

SUPPORTING GIRDER

4 x 4

ASPHALT ROOFING

12" MIN.**

2 x 4 PLATE

BAND

2 x 6 SILL

8" MIN.

*If square-edge 2•4•1 panels are used, blocking is required at unsupported edges.

**In areas of termite infestation or under conditions of adverse ground moisture, use 18" minimum.

Fig. 3-54　Glued Subfloor (Courtesy American Plywood Association)

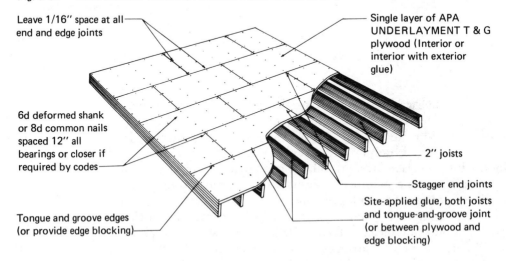

Leave 1/16" space at all end and edge joints

Single layer of APA UNDERLAYMENT T & G plywood (Interior or interior with exterior glue)

6d deformed shank or 8d common nails spaced 12" all bearings or closer if required by codes

2" joists

Stagger end joints

Site-applied glue, both joists and tongue-and-groove joint (or between plywood and edge blocking)

Tongue and groove edges (or provide edge blocking)

Cutting Holes and Notching Floor Joists

Cutting holes or notching floor joists is seldom done by carpenters, but it is often done by plumbers, electricians, and heating system installers. It is best that framing be done in a manner to avoid the need for cutting and notching. However, when notching is required it should be located only in the outer one-quarter of the joist span and should be limited to one-sixth the depth of the joist. Any holes which must be bored in the joists should not exceed 2″ in diameter, and the edge of the hole should not be closer than 2½″ to either the top or bottom edge of the joist under the greatest stress (see Fig. 3-55). Failure to follow these simple rules when cutting and notching results in unnecessarily weakened joists.

Fig. 3-55   Cutting Holes and Notching Joists

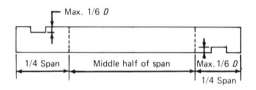

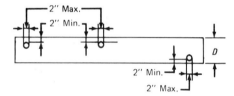

REVIEW QUESTIONS

1. What are the principal members of a wood floor frame?
2. Define a beam as used in house construction.
3. Outline the procedure for determining beam size.
4. What is the beam span?
5. What is the beam load width?
6. How do continuous joists affect beam load width?
7. What are some of the usual live loads and dead loads carried by the beam?
8. What factors are considered in final selection of beam size?

9. Find the total load carried by beams for the conditions listed.

SPAN	TOTAL LOAD POUNDS PER SQUARE FEET	BEAM LOAD WIDTH
6'	140	12'0"
7'	120	12'0"
8'	100	14'0"
9'	70	15'0"
10'	60	15'6"

10. How are beams installed?

11. What is beam bearing area?

12. What is the usual minimum bearing area required?

13. How is the beam height at the foundation wall adjusted?

14. Define a column.

15. What loads does a column supporting a beam carry?

16. Describe the installation of steel pipe columns.

17. Describe the installation of wood columns.

18. Define a joist.

19. What are the common joist spacings?

20. What two characteristics must a joist have?

21. What types of box sills are used in platform framing?

22. What type of box sill is used in balloon framing?

23. What are termite shields?

24. What is the minimum bearing distance for joists on concrete? On steel? On wood?

25. Why are joists framed into masonry walls given a fire cut?

26. What is the best way of framing joists at the beam?

27. How is the header joist or skirt layout started when joists are 16" on center?

28. What is a header joist? Tail joist? Trimmer joist?

29. Determine the load on a header joist 7' long supporting 10' long tail joists under usual conditions.

30. What is the purpose of bridging?

31. What is a subfloor?

32. Why are subfloor boards usually laid diagonally?

wall
construction

4

The tradesman has many names for the same item in house construction. Sometimes a slight change in the names will give it a completely new meaning, while at other times a completely different name is given to an item. Therefore the student should become familiar with the tradesman's terminologies.

WALLS AND PARTITIONS

Walls may be divided into categories according to the function they serve. Those walls which enclose the perimeter of the building are usually referred to as outside walls, outside partitions, or sometimes just as walls.

When the walls are used to divide the floor space into rooms they are called partitions. Partitions are divided into two categories according to the manner in which they are used. Bearing partitions run at right angles to the ceiling joists which rest on them. They are called bearing partitions because they support floor and ceiling loads. Partitions which run parallel to the joists do not support any loads and are called partitions or nonbearing partitions.

Outside walls and partitions are usually built of the same framing members and materials. The plate or sole is also known as a sole plate (see Fig. 4-1). This member, along with one of the top plates, will be marked off to show the location of all the studs, all the door and window openings, and all the intersecting partitions.

STUDDING

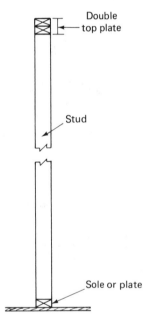

Double top plate

Stud

Sole or plate

Fig. 4-1   Selection Showing Parts of Typical Wall Frame

Studs are vertical members placed at close intervals along the length of the walls which act like columns in the wall to support the weight on the wall. The safe unbraced length for a wood column is 50 times its least dimension. Therefore the safe height for a 2 by 4 stud would appear to be between 6' and 7', but because the stud is braced by the sheathing and inside wall material, the least dimension becomes 3 ½" and the safe unbraced height becomes about 15'. Because the ceiling height in residential work seldom exceeds 8'0", there is no need to worry about the carrying capacity of 2 by 4 studs.

Lumber used for studding will vary with the locality in which work is being done. Often the material used is that which is readily available locally. Some of the more commonly used species are spruce, hemlock, white fir, and yellow pine. Red fir is also used for studding, but not as often as the other species mentioned. The grade of lumber used is usually Standard or better (No. 2 or better), and in some cases it may be Utility grade or No. 3. The latest lumber standard has established a "Stud" grade, and this is replacing other grades of lumber used for wall studs. Whatever grade of lumber is used, it must be suitable for studding and be approved by the local building code. It should be lumber that is straight, not twisted or bowed, and free from splits. Purchasing stud-grade lumber will usually eliminate lumber which is not suitable for studding with the exception that some stud lumber will be bowed or twisted. This material must not be used as studs or a crooked wall will be the result.

Stud Location

Common spacings for studding in residential construction are 16" O.C. and 24" O.C. The allowable spacing will vary with the various local building codes. The 16″ O.C. spacing is always acceptable for residential construction, and the 24″ O.C. spacing may be used in some localities if certain provisions of the code concerning sheathing material, bracing, and interior wall materials are complied with. The 24″ spacing is also acceptable for use in garages and temporary buildings.

By placing the regular studs 16″ or 24″ O.C. the carpenter is assured that there will be a stud every 48″ and will be able to use 48″ by 96″ panels with the joints between panels falling at the center of the studs.

Stud Length

The length of the regular studs may be determined by making a story pole. The story pole for studding is usually a straight 2 by 4 long enough to accommodate the finished ceiling height plus the thickness of all the framing lumber. For an 8′ ceiling the carpenter might select a 2 by 4 which is 10′ or 12′ long.

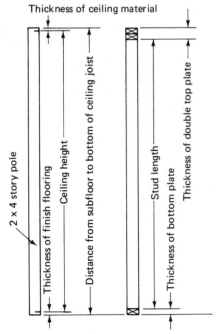

A. Initial layout   B. Completed layout   Fig. 4-2   Stud Length Story Pole

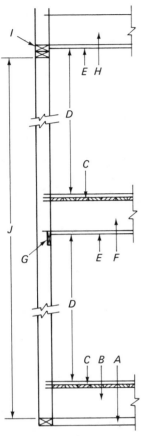

A - Sill plate
B - First floor joist
C - Sub- and finish floor
D - Ceiling height
E - Ceiling
F - Second floor joist
G - Ribbon layout
H - Ceiling joist
I - Double top plate
J - Stud length

Fig. 4-3   Story Pole Layout for Balloon Frame

To lay out the story pole for platform framing the carpenter first marks a line across the 2 by 4 near one end to represent the top of the subfloor (see Fig. 4-2A). Next the thickness of the finished floor is marked in and is followed by marks for the ceiling height and ceiling material thickness. The distance from the first to the last marks represents the distance from the top of the subfloor to the underside of the ceiling joists.

Before the stud length can be determined the thickness of the sole plate and top plates must be marked in (see Fig. 4-2B). The distance between the top of the bottom plate and the underside of the top plate or plates is the stud length.

If the building has an 8' 0" ceiling height, a ¾"-thick finish floor, and a ¾"-thick lath and plaster ceiling, the total distance from the top of the subfloor to the underside of the ceiling joists will be 97 ½". When the plan calls for a single bottom plate and a double top plate the combined thickness of three plates 1 ⅝" thick is 4 ⅞". The stud length will then be 97 ½" minus 4 ⅞" or 92 ⅝". The 92 ⅝" stud length has become very common in platform framing and as a result many lumber manufacturers cut 2 by 4's to an exact 92 ⅝" and stamp them, STUD—92 ⅝".

Under the new lumber standard 2" lumber is 1 ⁹⁄₁₆" or 1 ½" thick, depending on its moisture content. While the thickness of the 2 by 4 plates will affect the total height of the wall, the amount is small and the length of the standard stud is likely to remain 92 ⅝" for the foreseeable future.

To lay out a story pole for balloon framing the carpenter follows a procedure similar to that in platform framing. Because balloon framing usually involves a second story, the carpenter must select a straight 2 by 4 long enough to accommodate the thickness of the first- and second-floor construction and the two ceiling heights. A 2 by 4 that is 18' long is usually chosen as the stock length for these studs. The carpenter marks the location of the sill at one end of the story pole and proceeds to mark in the height of the first-floor joists, the flooring thickness, the ceiling height, and the ceiling thickness for the first floor (see Fig. 4-3). He has now located the bottom of the second-floor joist and may make the layout for the ribbon. After laying out for the ribbon the carpenter marks in the height of the second-floor joist, the second-floor thickness, the second-floor ceiling height and ceiling thickness, and has now located the underside of the ceiling joists. The layout for the top plates will complete the story pole layout for the stud length.

Additional layouts are made on the story pole for door and window openings for the purpose of determining the length of cripple studs. When buildings have various floor levels, stair landings, and other features which require a "standard" job reference, these too, will be marked on the story pole.

CORNER FRAMING

Outside corners of a building should be framed in a manner which will provide a straight corner, for good appearance, and nailing surfaces for the sheathing and siding on the outside and the interior wall material on the inside. Figure 4–4 shows two corner posts in plan view which will meet the requirements for good nailing surfaces at the outside and inside corners.

The 2 by 4's used to build the corner posts should be straight and free from defects which would weaken them or cause warping.

Corner posts are parts of the two intersecting walls. To build the corner post in Fig. 4–4A, the two 2 by 4's in the side wall are separated by three 2 by 4 blocks which are approximately 12″ long (see Fig. 4–5). Nailing the 2 by 4's to these blocks makes a rigid corner post to which the last stud in the end wall is nailed. Nailing the last stud in the end wall to the corner post in the side wall completes the corner and locks the two walls together.

To build the corner post in Fig. 4–4B, two straight 2 by 4's are nailed together to form an "L" (see Fig. 4–6). This corner

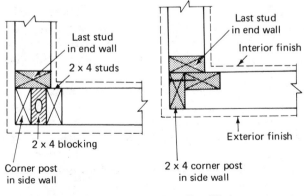

A. Standard corner post   B. Simplified corner post   Fig. 4-4   Corner Posts – Plan View

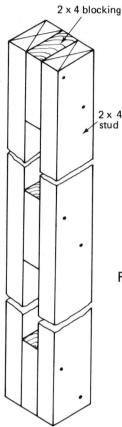

2 x 4 blocking

2 x 4 stud

Note: Studs nailed to 2 x 4 blocking with 16d nails

Fig. 4-5   Conventional Corner Post

post is placed in the side wall and provides a good nailing surface for the last stud in the end wall which completes the corner post.

There is some argument among carpenters as to which corner post is the best. Some say that post "A" is more rigid, and post "B" has a tendency to "split." If the two posts are compared it will be noted that each is made up of three 2 by 4 studs and each provides a good nailing surface for the sheathing on the outside and the interior wall material on the inside. Therefore both meet the purposes for which they were intended and can be considered equally good in that respect. Corner post "A" requires three short pieces of 2 by 4 which are readily available in the form of scrap on the job but takes slightly longer to make than corner post "B".

FRAMING WALL INTERSECTIONS

When walls intersect it is necessary to provide backing to which the interior wall material may be fastened. This backing can be made of extra 2 by 4 studs, or 1 by 4, 1 by 6, or 1 by 8 material nailed to the end stud with sufficient cross blocking to stiffen the backing. Figure 4-7 shows the plan view of a number of different ways to provide backing for intersecting walls.

Figure 4-7A shows a 1 by 6 or 1 by 8 nailed to the back of the end stud of the intersecting partition. This backing board is fastened with 8d nails spaced every 12" alternately near each edge of the stud. To provide stiffness and keep the partition stud in alignment, short 2 by 4's are nailed between the regular studs at the top, center, and bottom of the wall. These blocks are also fastened to the intersecting partition. Fastening the partitions together in this manner provides the stiffness necessary to keep cracking of plaster and even gypsum wallboard joints to a minimum at the corners.

In Fig. 4-7B a 1 by 4 was used as a backing member because the partition was placed alongside a regular stud, and the partition required backing on only one side. This 1 by 4 backing is nailed to the back of the end partition stud with 8d nails spaced every 12", and short 2 by 4's are installed at the top, center, and bottom of the wall to strengthen the corner.

To cut down on the time needed to install backing many carpenters install an extra stud on each side of the intersecting partition (see Fig. 4–7C). This method provides excellent backing, and to provide added stiffness a short 2 by 4 is nailed between the backing studs midway between the floor and ceiling. The end partition stud is nailed to this short 2 by 4 after the intersecting partition is erected.

The method of backing shown in Fig. 4–7D eliminates the short 2 by 4 of the previous method by having each of the backing studs overlap with the end stud of the intersecting partition by ⅜″. When this method is used the two walls are fastened together by nailing through the end stud at an angle into the backing studs with 16d nails as shown in Fig. 4–7D. Usually two 16d nails are sufficient.

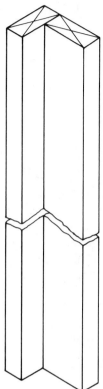

Note: 2 x 4's fastened
        together with
        16d nails
        spaced 12″ to
        16″ apart

Fig. 4–6   Simplified
Corner Post

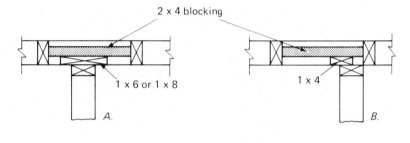

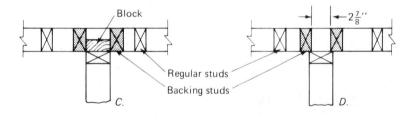

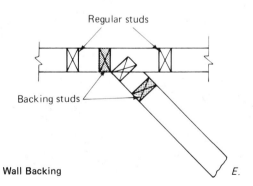

Fig. 4–7   Wall Backing

When partitions meet at an angle, backing is best provided by adding additional studs as required at the intersections. An example of this condition is given in Fig. 4–7E.

FRAMING DOOR AND WINDOW OPENINGS

When door and window openings are made in the wall it becomes necessary to remove some of the regular studs. To support the framework above the opening, special framing members are installed. For doors these framing members include headers, cripple studs, shoulder studs, and door studs (see Fig. 4–8).

In wall construction, a header is a beam above a door or window opening which supports the weight of construction and floor loads above the opening. Cripple studs are the short studs which fill the space between the header or sill and the top or bottom plate, respectively. Cripple studs are placed on the regular stud spacing.

Shoulder studs form a ledge or shoulder to support the header. They are fastened to the regular studs, which are nailed to the headers.

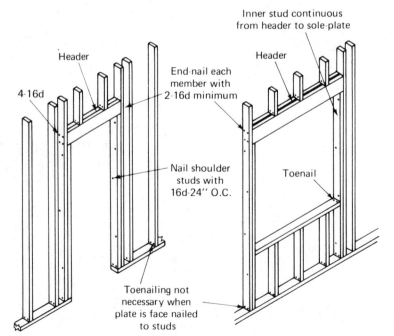

Fig. 4–8   Typical Door and Window Openings

In some localities shoulder studs are known as cripples or trimmers. In addition to the above, window openings also require a rough sill and cripple studs between the rough sill and the sole plate.

Door Openings

To frame a rough opening for a door the carpenter must know the rough opening size. This rough opening is the sum of the door size, the thickness of the frame on each side of the door, plus ¼″ to ½″ framing allowance on each side of the frame. For interior door openings a ¼″ framing allowance is made on each side of the door jamb to take up irregularities between the rough framing lumber and the finished door frame.

The rough opening for a 3′ 0″-wide door can be found by adding all the necessary items as follows:

Door width	36″	
Frame thickness	1 ½″	(¾″ each side)
Framing allowance	1″	(½″ each side)
	38 ½″	Rough Opening Width

The plate layout for this opening is shown in Fig. 4–9. Notice that the regular 16″ O.C. marks are carried straight through and are not altered because of door layout. The center of the opening is located on the plate in accordance with dimensions given on the plan, and one-half the rough opening distance is marked off in each direction. After the size of the opening is marked the shoulder studs are drawn in. Notice that they are indicated by O's rather than X's. An additional regular stud is placed alongside the shoulder stud and is known as a "door stud." This door stud is the one to which the header is nailed.

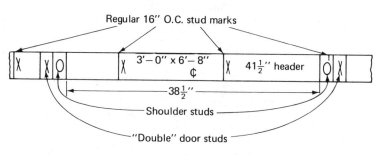

Regular 16″ O.C. stud marks

3′ − 0″ × 6′ − 8″ ₵   41½″ header

38½″

Shoulder studs

"Double" door studs

Fig. 4-9   Plate Layout for 36″ Wide Door

In addition to marking the location of the framing members, the carpenter often marks in the size of the door and the length of the header. Because the header rests on the shoulder studs, 3″ (1½″ on each side) must be added to the rough opening size to get the header length. The header in this example must be 41½″ long. Marking the header length on the plate in this manner saves time in cutting and fabricating, because the workmen do not have to take out their rules and measure each opening as they come to it.

Header Size

The size of header needed will vary with the width of the rough opening. Local building codes have requirements concerning header size, and a check of the various codes will show that there are variations in the requirements among the different codes. Therefore the carpenter must know the building code requirements for the areas in which he works.

Table 4-1 gives the maximum spans for wall headers in outside walls and bearing partitions. These spans are acceptable as a general rule, and where there are no other code requirements they can be used with confidence.

The span of the header is the distance between the shoulder studs and is equal to the rough opening width. When openings larger than those given in Table 4-1 are needed, the double 2 by 12 header may be reinforced with a steel plate placed between the 2 by 12's. The double 2 by 4, 2 by 6, and 2 by 8 headers are sometimes replaced by single 4 by 4, 4 by 6, and 4 by 8 headers respectively.

When a double 2″ header is used its combined thickness is only 3″ (1 ½″ + 1 ½″), but the 2 by 4 wall is 3 ½″ wide. There-

Table 4-1. Maximum spans for wall headers, outside walls, and bearing partitions

HEADER SIZE	MAXIMUM SPAN
2 — 2 by 4 on edge	4′0″
2 — 2 by 6 on edge	5′6″
2 — 2 by 8 on edge	7′6″
2 — 2 by 10 on edge	9′0″
2 — 2 by 12 on edge	11′0″

fore the ½″· difference must be made up by a filler strip. This filler strip is usually made from wood lath, ½″ plywood, or specially made rippings. The filler is usually placed between the header members, and 16d nails are driven through the headers and filler strips to form the header. The header is nailed together in this manner so that it will act as a unit (one piece) when it is installed in the wall.

An alternative method of building the added thickness is to place the filler strips on the outside of the header after the header members are nailed together (see Fig. 4–10). This method is said to provide added header stiffness because of the friction between the header members.

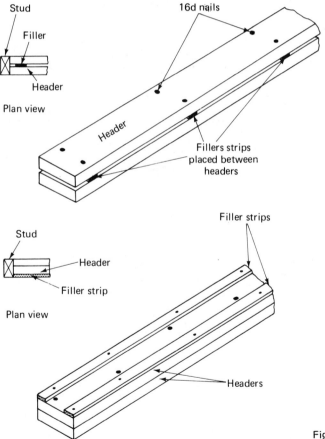

Fig. 4–10   Wall Headers with Filler Strips

Rough Opening Height

The distance from the subfloor to the bottom of the wall header is the same for most door and window openings and is governed by the height of the door frame.

Determination of the height of the opening can be calculated mathematically, or it can be made by additional layouts on the story pole. To lay out the height on the story pole the length of the door opening in the frame is first marked off on the story pole (see Fig. 4-11). For a standard 6'8" door this distance is 80½". This allows 80" for the door and ½" for the threshold which is placed below the door. Next the thickness of the head jamb of the door frame is marked in. Head jambs are usually ¾" thick. Finally a framing allowance of ½" is marked

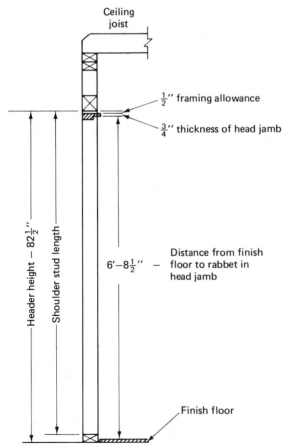

Fig. 4-11   Locating Header Height on Story Pole

in, and the location of the bottom of the wall header has been established.

In checking the layout, it will be found that in this case the distance from the subfloor to the header is 82½". The 82½" includes a ¾" allowance for the thickness of the finished floor. In most cases this is also the distance from the floor to the bottom of the window headers. The length of the shoulder stud which runs from the bottom plate to the header can be found by subtracting the plate thickness from 82½". Using a 1½"-thick plate we would find the shoulder studs to be 81" long.

Now that the location of the bottom of the header has been established the height of the header can be drawn in to determine the length of the cripple studs above the header (see Fig. 4-12). Notice that each different lumber size of header is drawn in and each line is labeled. The length of the cripples or blocks, as they are sometimes called, may also be marked on the story pole.

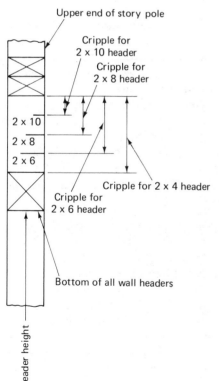

Fig. 4-12   Story Pole Layout for Header Cripple Studs

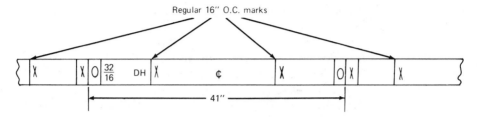

Fig. 4-13   Plate Layout for Window Opening

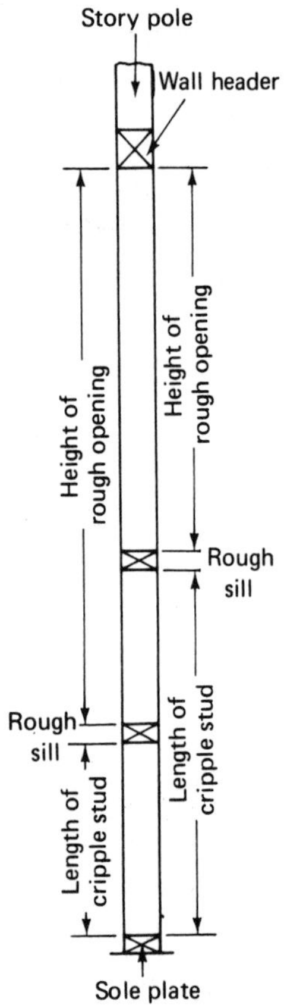

Window Opening

The layout of the rough opening for a window is similar to that for a door. The carpenter must know the size of the frame and add a ¼″ to ½″ framing allowance on each side to get the rough opening size. When standard double-hung windows are used the carpenter may add 6″ or 6 ½″ to the glass size to get the width of the rough opening and 9″ to 9 ½″ to the glass size to get the height of the rough opening.

Because of the wide variety of window frames available and the varying construction of these frames, it is recommended that the carpenter obtain catalogs from the manufacturer whose frames he is using. These catalogs list the size of the frames and the rough opening needed for each different glass size. Reference to these catalogs can prevent making mistakes in rough opening size and save the carpenter a lot of extra unproductive work.

The first step in the layout of the rough opening is to locate the window in accordance with the dimensions on the plan and to mark off the required width in the same manner as for a door opening. Figure 4–13 shows a typical plate layout for a window. Notice that the carpenter has marked in the type of window, which is a 32″ by 16″ double-hung unit, and the length of the header, which is 41″.

Window Cripple Studs.   The length of the window cripple studs may be determined by making additional layouts on the story pole. Starting at the bottom of the header the carpenter measures down a distance equal to the rough opening height for the window frame and places a mark on the story pole (see Fig. 4–14). Next the thickness of the rough sill is drawn in and the length of the cripple studs may be measured off between the bottom plate and the rough sill.

When there are many different window heights drawn in on the story pole, each is labeled, and the length of the cripple studs are marked in for future reference.

Note: Story pole may have layouts for many different sizes of windows — each layc should be indentifiec

Fig. 4-14   Story Pole Layout – Cripple Studs

Make Openings in Balloon Framing

Openings in balloon framing are made in a manner similar to that for platform framing, the main differences being that the shoulder studs run down to the sill and that the second-

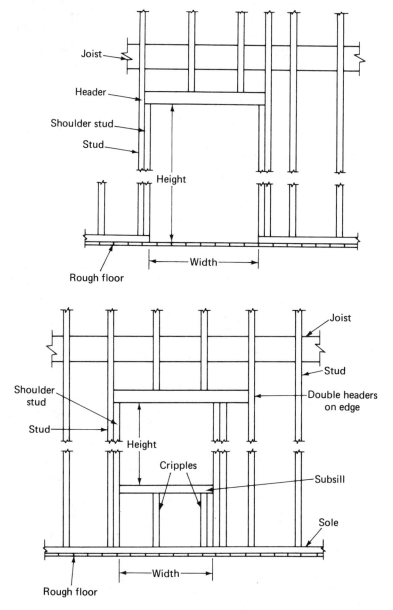

Fig. 4-15   Methods of Framing Openings in Balloon Frame

floor openings must be considered when making the layout. In many balloon-frame structures the windows on the second floor are in line with those on the first floor and are the same size as the windows with which they line up. This type of design simplifies construction (see Fig. 4–15).

The story pole for the openings in a balloon-frame building is marked off in a manner similar to that for platform framing. Additional marks are placed on the stud-length story pole to indicate the floor lines, door heights, header depths, the height of window rough openings, and rough sills (see Fig. 4–16).

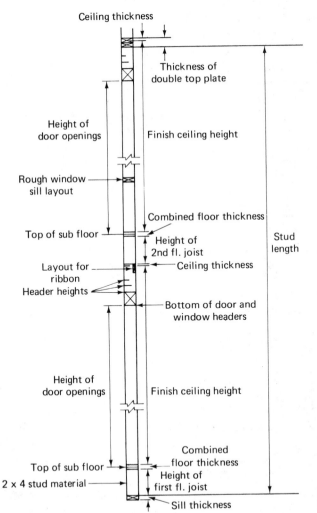

Ceiling thickness

Thickness of double top plate

Height of door openings

Finish ceiling height

Rough window sill layout

Top of sub floor

Combined floor thickness

Height of 2nd fl. joist

Ceiling thickness

Layout for ribbon

Header heights

Bottom of door and window headers

Stud length

Height of door openings

Finish ceiling height

Top of sub floor

Combined floor thickness

2 x 4 stud material

Height of first fl. joist

Sill thickness

Fig. 4-16   Story Pole – Locating Wall Headers in Balloon Frame

LAYOUT OF WALL PLATES

The location of all regular studs, corner posts, door and window openings, and intersecting partitions and backing studs is marked off on the top and bottom plates. Almost always the layout is completed for the outside walls before layout is begun on the partitions, but the procedure followed will be the same for all walls and partitions.

Marks are made on the floor which establish the inside of the 2 by 4 wall, and chalk lines are snapped on the deck (subfloor) between the marks. This is the line along which the plate will be nailed.

When the wall runs at right angles to the joists the studs should be placed directly above the joists or directly at the edge

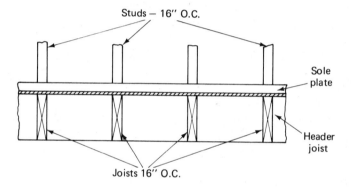

Studs — 16″ O.C.

Sole plate

Header joist

Joists 16″ O.C.

a. Studs placed over joists

Note: Placing studs in the manner
illustrated makes provision
for placement of heating pipes
between joists and studs

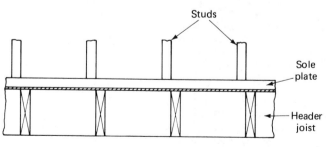

Studs

Sole plate

Header joist

b. Studs placed along side of joists

Fig. 4-17   Placement of Studs and Joists

of the joists (see Fig. 4–17). Studs which are placed directly over the joists have full support of the joists, and studs which are placed at the edge of the joists get sufficient support. An important factor to remember is that when studs are placed in this manner there always is sufficient room for heating pipes placed between the joists to be turned up and run into the walls where required.

Using straight 2 by 4's for plates, the carpenter starts his layout so that the regular studs are positioned at the joists as discussed above. If the walls are erected on a concrete floor he starts his layout by placing a corner post at the end of the wall and measuring 15¼″ to the side of the first stud (see Fig. 4-18). By starting the layout in this manner and continuing 16″ O.C. from the first mark, the distance from the end of the wall to the center of the third stud from the corner post will be 48″. This type of layout makes it possible to use 48″ by 96″ sheets of sheathing and have all the end joints fall at the center of the stud.

Following the layout for all the regularly spaced studs, the intersecting walls are located in accordance with the plans and the backing studs are marked in. The location of the doors and windows is now marked in according to the dimensions on the plan, and the rough openings, shoulder studs, and double studs are also drawn in.

Gaining Space

If it is necessary to conserve on floor space, partitions are often built by placing the studs flat in the wall. When this is done a 2 by 2 is used for the plates, and an additional 2″ is gained in room size (see Fig. 4–19).

When more space is needed within the wall to make room for large pipes, 2 by 6 plates and studs are used. It is important that the studs be located directly above the joists so that there

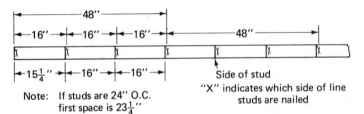

Note: If studs are 24″ O.C.
first space is 23¼″

Side of stud
"X" indicates which side of line
studs are nailed

Fig. 4-18   Starting Plate Layout

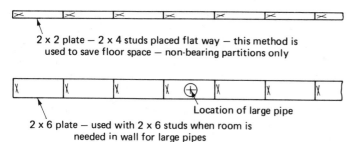

2 x 2 plate — 2 x 4 studs placed flat way — this method is
used to save floor space — non-bearing partitions only

Location of large pipe

2 x 6 plate — used with 2 x 6 studs when room is
needed in wall for large pipes

Fig. 4-19   Special Walls — Plan
View

will be ample room for the pipes to run through the floor. Placing these studs directly above the joists also helps to avoid the
need for extra cutting and joist framing when the plumbing and
heating pipes are installed.

BUILDING THE WALL

The following discussion relates to platform framing, although
much of the procedure discussed can also be applied to other
types of framing.

Fig. 4-20   Assembling
a Wall

After the plates have been laid out the studs are placed on the floor between the top and bottom plates approximately at markings on the plates. These studs are later aligned with the marks on the plates and fastened to the plates with two 16d nails driven through the plates at each end of the stud (see Fig. 4-20).

Window headers are installed according to the plate layout. Cripple studs are toenailed to the header at the location indicated on the plates with four 8d nails each, and the plate is fastened to the cripple studs with two 16d nails each in the same manner as with regular studs (see Fig. 4-21). Each end of the header is nailed to a window stud (see Fig. 4-8) with four or more 16d nails. For ease in nailing, the header and window studs should be fastened together before the regular studs adjacent to the opening are installed.

After assembling the wall framework, the top plates are doubled. An opening slightly wider than the width of the intersecting wall plates is left for each intersecting wall marked on the top plate. This is done to provide for the interlocking of all intersecting walls (see Fig. 4-22). The double top plate is also used to bridge the joints in the top plate and thereby stiffen it. This aids in keeping the plate aligned. The straightest 2 by 4's available on the job should be selected for plates, because straight plates will automatically lead to straight walls.

The double top plates on the end walls (the walls which run between the side walls) must be left longer than the wall so that they will overlap with the side walls and provide a good interlock at the corner (see Fig. 4-22). To make erection easier, the section of plate which projects beyond the end of the wall is usually left off until the end wall is erected. This piece, which

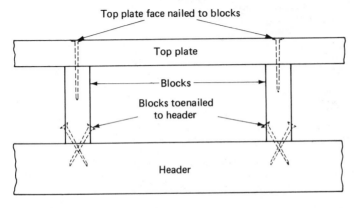

Top plate face nailed to blocks

Top plate

Blocks

Blocks toenailed
to header

Header

Fig. 4-21   Fastening   Blocks   to Header

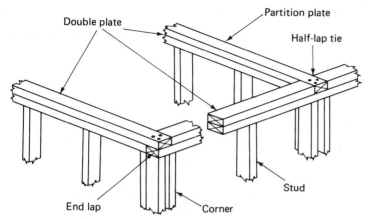

Fig. 4-22   Intersecting Top Plates

is usually about 4′ long, is nailed into place as soon as the wall is erected.

Often before the wall is erected corner braces and sheathing are installed, but before this can be done the wall must be straightened and squared.

Straightening is accomplished by aligning the bottom plate with the chalk line snapped on the floor for the purpose of locating the wall. As the edge of the plate is aligned with the chalk line it is held in place by toenailing it to the subfloor.

Fig. 4-23   Squaring a Wall (Courtesy National Forest Products Association)

The wall is squared by measuring the diagonals and moving the top plate as required to square the wall. When the diagonal distances are equal the wall is square (see Fig. 4–23).

WIND BRACING

Walls must be braced against lateral loads. The most common lateral load is caused by the wind. Other lateral loads are induced by objects and persons leaning against the wall. Bracing against lateral loads is obtained in a number of ways and is commonly referred to as wind bracing.

Board sheathing applied horizontally or diagonally provides varying amounts of bracing (see Fig. 4–24 and 4–25). Other types of bracing are the cut-in 2 by 4 (Fig. 4–26), the let-in 1 by 4 (Fig. 4–27), 48″ by 96″ composition panels and plywood sheets, herringbone bracing, and steel strapping.

The amount of rigidity the various types of bracing impart to a wall varies with the type and method of installation. Rigidity of the various types of bracing is compared in Table 4–2. The standard wall panel to which all types of bracing are compared was 8′ by 8′ with studs 16″ O.C. and sheathed with sea-

Fig. 4-24 Horizontal Board Sheathing (Courtesy Western Wood Products Association)

Fig. 4-25   Diagonal Sheathing (Courtesy Western Wood Products Association)

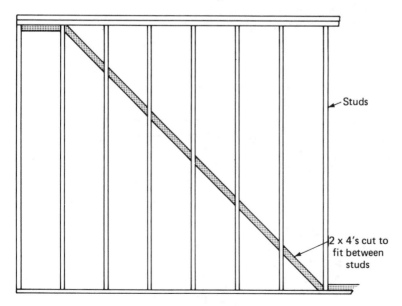

Studs

2 x 4's cut to fit between studs

Fig. 4-26   Cut in 2 by 4 Braces

Fig. 4-27   Let in 1 by 4 Brace

soned 1 by 8 material, applied horizontally and nailed with two 8d common nails at each stud crossing. This panel did not have any door or window openings.

A panel sheathed with green 1 by 8 material and allowed to stand for 30 days before testing showed a 40% reduction in rigidity when tested. This test proved the advantages of using seasoned lumber for sheathing to gain strength and also a wall that was more resistant to air infiltration. As the green material seasoned, spaces opened between the boards. These spaces were responsible for the loss in rigidity.

Herringbone bridging (Fig. 4-28) placed in a wall with horizontal sheathing showed a slight increase in rigidity. However, compared to the cost of installation this type of bracing adds little to the rigidity of the wall and can safely be omitted.

Cut-in 2 by 4 braces (Fig. 4-26) used with horizontal

Table 4-2. Relative rigidity of frame construction based on tests at Forest Product Laboratory

PANELS WITHOUT DOOR AND WINDOW OPENINGS

PANEL CONSTRUCTION			(scale: 100 200 300 400 500 600 700)
Nails	**Sheathing**	**Bracing**	
2 8d	Horizontal 1 by 8 seasoned	none	
2 8d	Horizontal 1 by 8 green*	none	
2 8d	Horizontal 1 by 8 seasoned	herringbone bridging	
2 8d	Horizontal 1 by 8 seasoned	2 by 4 cut-in braces	
2 8d	Horizontal 1 by 8 seasoned	1 by 4 let-in braces	
2 8d	Horizontal 1 by 8 seasoned	steel strap	
2 8d	Diagonal 1 by 8 seasoned	none	
2 8d	Diagonal 1 by 8 green*	none	
	No Sheathing—Wood Lath, and Plaster		
6d**	$\frac{1}{4}$" plywood	none	
8d†	$\frac{25}{32}$" fiberboard	none	
1$\frac{3}{4}$" roof†	$\frac{1}{2}$" fiberboard (MD)	none	
8d‡	$\frac{5}{8}$" plywood	none	
2 8d	No sheathing	1 by 4 let-in braces	

PANELS WITH DOOR AND WINDOW OPENINGS

Nails	**Sheathing**	**Bracing**	
2 8d	Horizontal 1 by 8 seasoned	none	
2 8d	Horizontal 1 by 8 seasoned	1 by 4 let-in braces	
2 8d	Diagonal 1 by 8 seasoned	none	
6d**	$\frac{1}{4}$" plywood	none	
8d†	$\frac{25}{32}$" fiberboard	none	
	No sheathing—Wood, Lath, and Plaster		

*Panel allowed to stand 30 days.
**Nail spacing: 5" at perimeter, 10" intermediate.
†Nail spacing: 3" at perimeter, 6" intermediate.
‡Nail spacing: 6" at perimeter, 12" intermediate.

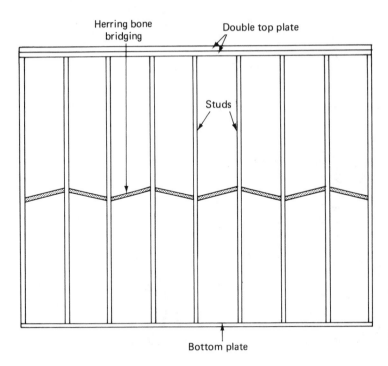

Herring bone bridging

Double top plate

Studs

Bottom plate

Fig. 4-28   Herringbone Bridging

sheathing increase the rigidity of the wall about 60%. The amount of increased rigidity is partially dependent on how well the miter cuts fit the studs and on how well the 2 by 4 braces are nailed into place.

The let-in 1 by 4 brace used with or without sheathing shows an increase in rigidity of 150% over the base panel. Therefore it may be concluded that the let-in 1 by 4 is the best type of bracing to use with all types of sheathing.

A steel strap 1″ wide by $\frac{1}{32}$″ thick nailed diagonally over the sheathing increases the rigidity of the wall by about 90%.

Diagonally applied board sheathing increases the rigidity of a wall with no door or window openings to about four times that of horizontal sheathing on a wall with no bracing. However, with the introduction of wall openings the increase is only about 20% over the standard panel.

The newer materials which are replacing board sheathing have bracing characteristics much better than boards. Plywood far exceeds the bracing of boards when nailed according to the recommendations of the American Plywood Association. Fiberboard also provides better bracing than board sheathing when nailed in accordance with the manufacturer's recommendations.

As indicated in Table 4-2, plywood and fiberboard sheathing can provide adequate bracing, even in walls with door and window openings. While these materials provide adequate bracing many building codes still require a wind brace. When a wind brace is required, the let-in 1 by 4 is the best choice.

Let-in 1 by 4 braces are located near the ends of the wall. The 1 by 4 is placed on the face of the studs at an angle of approximately 45° and may be held or tacked (nailed temporarily) in place while the carpenter marks its location on each stud and plate. The recess is made by cutting into the studs and plates with a portable power saw set to the thickness of the 1 by 4. Several cuts are made between the layout lines, and the narrow strips of wood are broken off with a hammer (see Fig. 4-29). After the recesses have been made in the studs the 1 by 4 is nailed at each plate and stud crossing with two 8d nails.

In recent years, where allowed by building codes, many builders have been using a 4' by 8' sheet of plywood at each end of the wall in place of other kinds of bracing. This procedure saves time, reduces building cost, and results in a wall rigidly braced against wind loads (see Fig. 4-30). When ½" thick fiberboard sheathing is used, ½" thick plywood bracing provides a

Fig. 4-29   Cutting and Chipping
Recess for 1 by 4 Brace

Fig. 4-30   Plywood Sheet Wind Brace

uniform wall surface. If ¾″ thick fiberboard is used with ½″ plywood bracing, furring strips must be applied before installing the finished siding. The alternative is to use plywood bracing material which is the same thickness as the sheathing material.

SHEATHING

Wall sheathing serves to enclose the building, reduce infiltration, brace the wall, and provide a base for the siding. Different types of sheathing have different insulating and different bracing capabilities as well as different characteristics as a base for siding.

Wood boards are the oldest type of sheathing in use. Most boards used for sheathing are either 1 by 6 or 1 by 8. Although wider boards are sometimes used, they should be avoided because the greater amount of shrinkage that takes place in wider boards leads to gaps or spaces between the boards. These boards may be square edged, but usually they are either tongue and groove or shiplap (see Fig. 4-31). Tongue and groove and shiplap are preferred over square edged boards because the edges fit

Shiplap

Tongue
and
groove

S4S
(square edge)

Fig. 4-31   Types of
Board Sheathing

together to reduce the amount of air and heat which can infiltrate the wall.

A large number of lumber species are used for wood sheathing. Therefore the kind and grade of lumber used for sheathing will vary in different geographical areas with availability and local practice. Some of the species used to make sheathing lumber are southern yellow pine, spruce, hemlock, white fir, and various western pines.

As indicated in the section on wind bracing (p. 112), boards may be applied either horizontally or diagonally. All end joints should occur at the center of a stud and should be double nailed. The boards should be nailed with at least two 8d nails at each stud crossing.

Composition or fiberboard sheathing made from sugar cane, corn stalks, wood pulp, and other materials is available under various trade names. It is available in sheets ½″ thick and ²⁵⁄₃₂″ thick. The standard-size ½″-thick sheet is 48″ by 96″. Some manufacturers offer larger-size sheets, but these are not used extensively.

Composition sheathing ²⁵⁄₃₂″ thick is available in 48″ by 96″ square-edged sheets and 24″ by 96″ sheets with V-groove matching on the long edges.

The 48″ by 96″ panels are usually installed vertically. This allows the top and bottom edges to be secured to the plates. The remaining edges are placed at the center of the studs.

The 24″ by 96″ panels are installed horizontally. All end joints fall at the center of a stud. The joints in adjacent rows of sheathing should be arranged so that the ends of the sheets do not fall on the same stud. The V-groove edge joints between sheets of adjacent rows fit together to prevent air infiltration and also to keep the sheets aligned.

Nails or fasteners used for composition sheathing should be of a type and size recommended by the manufacturer. Using fasteners smaller than those recommended seriously reduces the strength of the wall if the sheathing is being used for wind bracing. It is also important that a sufficient number of fasteners be used as recommended in the sheathing manufacturer's installation instructions.

Gypsum sheathing is manufactured in 24″ by 96″ sheets ½″ thick. The face and back paper of gypsum sheathing is black and is treated with a water-repellent material. Gypsum sheathing is also available with aluminum foil on one side. Sheathing with foil is usually used on garages where the studs are 24″ O.C. The sheets are applied vertically with all joints at the center of the stud. This application gives an inside appearance of one flat sheet of aluminum and provides a bright interior.

The long edges of gypsum sheathing are matched with a V groove. Regular gypsum sheathing is always applied horizontally, with the end joints occurring at the center of a stud. Roofing nails 1 ½" to 2" long are usually used to fasten gypsum sheathing to the studs. They are spaced every 6" along each stud. Unlike other sheathing materials, which must be cut with a saw, gypsum sheathing is cut by scoring the face paper with a knife and breaking the core by bending the sheet at the score mark.

Plywood sheathing is manufactured in thicknesses of 5/16", 3/8", ½", 5/8", ¾", and 7/8" and in different grades. The thickness of sheathing which can be used in a given locality is governed by the local building code. Other factors which govern the thickness that can be used are the grade of plywood and the spacing of the studs. Of the thicknesses available 5/16", 3/8", and ½" are the most commonly used for sheathing. The stud spacing and nailing recommendations of the American Plywood Association are given in Table 4-3. The Panel Identification Index in column 1 of Table 4-3 refers to the rafter/joist spacing assigned to the different panel grades.

A complete listing of plywood grades and characteristics can be found in the American Plywood Association publications listed in the bibliography.

Plywood sheathing is usually installed with the face grain running vertically. When this is done on walls which are less

Table 4-3. American Plywood Association nailing recommendations for plywood sheathing.

Plywood Wall Sheathing*
(Plywood continuous over two or more spans)

Panel Identification Index	Panel Thickness (inches)	Maximum Stud Spacing (inches) Exterior Covering Nailed to:		Nail Size**	Nail Spacing (inches) Panel Edges (when over framing)	Intermediate (each stud)
		Stud	Sheathing			
12/0, 16/0, 20/0,	$\frac{5}{16}$	16	16†	6d	6	12
16/0, 20/0, 24/0	$\frac{3}{8}$	24	16 24†	6d	6	12
24/0, 32/16	$\frac{1}{2}$	24	24	6d	6	12

*When plywood sheathing is used, building paper and diagonal wall bracing can be omitted.

**Common smooth, annular, spiral thread, or T-nails of the same diameter as common nails (0.113" dia. for 6d) may be used. Staples also permitted at reduced spacing.

†When sidings such as shingles are nailed only to the plywood sheathing, apply plywood with face grain across studs.

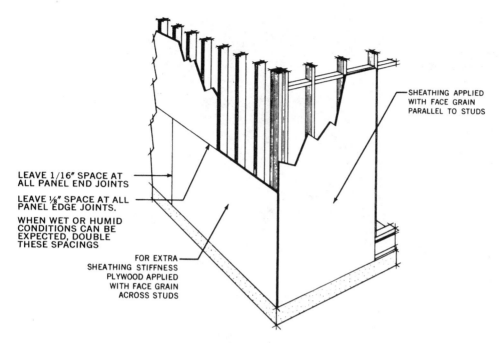

SHEATHING APPLIED
WITH FACE GRAIN
PARALLEL TO STUDS

LEAVE 1/16″ SPACE AT
ALL PANEL END JOINTS

LEAVE ⅛″ SPACE AT ALL
PANEL EDGE JOINTS.
WHEN WET OR HUMID
CONDITIONS CAN BE
EXPECTED, DOUBLE
THESE SPACINGS

FOR EXTRA
SHEATHING STIFFNESS
PLYWOOD APPLIED
WITH FACE GRAIN
ACROSS STUDS

Fig. 4-32   Plywood Sheathing (Courtesy American Plywood Association)

than 8′ high all edges of the sheet are supported either by the
plates or by the studding. If the walls are higher than the length
of the plywood sheet, blocking must be installed between the
studs to support the end edge of the sheet (see Fig. 4-32). When
plywood sheathing is installed with the face grain running at
right angles to the studs no edge blocking is necessary.

The omission of edge blocking is possible when the face
grain runs at right angles to the studs, because the plywood is
stiffer in the direction parallel to the grain. This is due to the
three plys running in that direction. There are only two plys
running at right angles to the face grain, and this makes the
sheet less rigid in that direction, making it necessary to support
the plywood at the ends.

FABRICATION AND ERECTION

While the wall plates are being laid out, one member of the
carpentry crew is usually busy cutting headers, blocks, cripple
studs, shoulder studs, and regular studs if they are not pur-

Fig. 4-33   Wall Fabricating

chased ready cut. Depending on the operation he will do his cutting either with a portable power saw, or a radial-arm builder's saw. As the lumber is cut it is piled neatly in a place where other crew members may easily pick it up and carry it to point of need.

The procedure for building a wall one story in height utilizing the tip-up method of house construction is outlined in the following paragraphs (see Fig. 4-33, 4-34 and 4-35).

Plate Layout

The location of all regular studs, intersecting walls, backing, and door and window openings is marked on the top and bottom plates in the manner previously outlined in this chapter on pp. 96-108.

Locating Materials

Regular studs are placed between the top and bottom plates at each stud mark. Shoulder studs are placed at each door and window opening along with proper headers, cripple studs, blocking, and rough sill as needed. Corner posts are placed at ends of outside walls where needed.

Fabricating the Framework

Studs are nailed to the top and bottom plates with two 16d nails driven through the plates into each end of the studs. Shoulder studs are nailed with 16d nails to the door or window stud at the side of the opening.

Spacers are placed between the headers as needed, and the headers are fastened together with 16d nails. The location of the cripple studs or blocks is transferred to the header from the top plate, and cripple studs (blocks) are toenailed to the header at the locating marks. The header is now placed atop the shoulder studs and fastened to the door or window stud with four or more 16d nails at each end, and the cripple studs (blocks) are fastened to the top plate in the same manner as studs. Cripple studs are nailed to the bottom plate where required, and the rough sill is marked with the location of the cripple studs at the bottom plate and nailed to the top of each cripple stud with two 16d nails. The rough sill is then toenailed to the shoulder studs with either 8d or 16d nails.

Studs for backing are nailed to the plates as indicated by the markings on the plates. Corner posts are fastened in the same manner as studs at the proper locations.

After all the framing members are nailed together the top plate is doubled, with spaces left for the intersecting walls. Outside walls usually require wind bracing and sheathing, but before these can be installed the wall must be aligned and squared.

Aligning and Squaring

The assembled wall is aligned by placing the bottom plate along the chalk line which has been marked on the floor to locate the wall. As the plate is aligned with the chalk line it may be held in place by toenailing it to the wood floor. When the bottom plate has been aligned the top plate will also be aligned if it is nailed securely to the studs. A quick check of the alignment at the top plate will reveal any studs which are not nailed securely. After these are hammered down the top and bottom plates are straight.

The wall is checked for squareness by measuring the diagonals. If they are not equal the wall is moved at the top plate in the direction required to make the diagonals equal. Following this operation the wind bracing and sheathing are installed.

Wind Bracing and Sheathing

Bracing is installed as described on p. 117. The sheathing is installed over the studs with all end joints occurring at the center of a stud. Window and door openings are marked and cut out as the application of sheathing progresses. As the sheathing is applied it is tacked to hold it in place and is nailed solidly after all the sheathing is in place.

Wall Erection

The completed wall may be erected by means of a mechanical jack, or it may be raised by hand. If jacks are used, two men can easily raise a wall 50′ long (see Fig. 4–34). Long walls raised by hand will require more men, and the number of men needed will be determined by the size of the wall and the type of sheathing on it. Heavy walls require more manpower to lift

Fig. 4–34   Erecting Wall with Wall Jacks (Courtesy Proctor Products, Inc.)

Fig. 4-35   Erecting Wall by Hand

them (see Fig. 4-35). The size of the lifting crew should be big enough so that none of the crew is injured as a result of lifting more than his capacity.

Using Wall Jacks

After the wall is completed and ready for erection, the wall jacks are fastened to the wall and floor. While the procedure can be varied somewhat, it is recommended that the lifting cables be attached to the wall at a distance of one-quarter the length of the wall from each end. Wall stops on the jacks are adjusted for the height of the wall, and the jacks are fastened to the floor flush against the lifting bracket (see Fig. 4-36). As lifting begins the jacks are in a vertical position, but as the wall is raised the jacks remain in contact with the top of the wall. When a vertical position is reached the wall is held firmly by the wall stops, and the wall jacks provide a substantial brace until another means of bracing is accomplished.

As the walls are erected they are fastened to the floor in line with the layout marks and are held erect with a number of temporary braces. The number of braces used should be sufficient to hold the wall against wind load until intersecting walls are erected.

Fig. 4-36   Using Wall Jacks (Courtesy Proctor Products, Inc.)

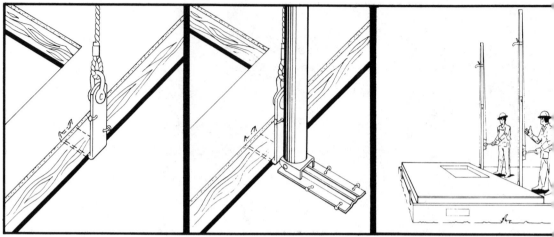

The lifting brackets are placed under the top plate or header and securely nailed.

The Jacks are placed upright, flush against the lifting bracket. The hinged foot is securely nailed to the sub-floor near a floor joist.

The lift begins, both Jacks being operated in unison.

A. If one man is raising the wall, he simply moves from one Jack to the other avoiding excessive twisting of the wall.

The Jacks remain in contact with the top edge of the wall during the entire lift.

When the wall reaches the upright position, it is stopped and held firmly by the wall stops.

The Wall Jacks provide a substantial brace until the wall can be permanently braced with Proctor Adjustable Wall Braces.

After all the walls have been erected and the intersecting plates fastened together, it is necessary to straighten the top of the walls and temporarily brace them in place. This aligning can be done by stretching a string over spacer blocks from one end of the building to the other, or by sighting along the top of each wall.

As one member of the crew sights the wall either with or without the aid of a taut string, he or she instructs other members of the crew to adjust braces that either push the wall out or pull it in. When the wall appears properly aligned the temporary braces are fastened in place.

When using 2 by 4's for bracing it is common practice to nail the upper end of the brace to the wall first and to fasten the lower end when alignment is achieved. These braces are left in place until the roof sheathing is nailed in place. The braces are then removed, and the lumber is used for miscellaneous framing within the building.

Some builders use adjustable wall braces in place of 2 by 4's. These braces have an adjusting screw at one end. Initially the brace is nailed to the top of the wall and the floor. Then as the wall is sighted the screw is adjusted to bring the wall into alignment (see Fig. 4–37).

Fig. 4-37   Adjustable Wall Braces (Courtesy Proctor Products, Inc.)

¾" Fast Thread

1" Square Tubing - Cadmium Plated

REMOVE WITH PRY BAR OR HAMMER

These braces also remain in place until the roof sheathing is installed. They are easily removed and moved to the next job.

MISCELLANEOUS FRAMING

There are many framing jobs connected with wall framing that have not yet been discussed. Some of the more common framing jobs will be discussed in the following paragraphs.

Ceiling Backing

When partitions run parallel to the ceiling joists it becomes necessary to provide backing at the top of the wall for fastening the ceiling material (see Fig. 4-38). This backing usually consists of a 1 by 8 board which is nailed to the top of the wall and projects on each side of the wall. If backing is needed on only one side of the wall because of the placement of the ceiling joists, a 1 by 4 is usually used. The backing is braced with 2 by 4 blocking installed between the joists and nailed to the

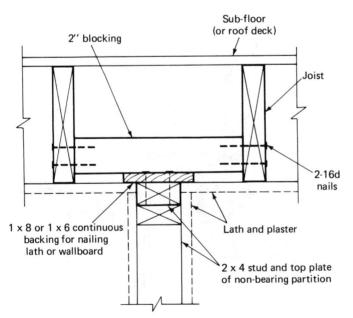

Fig. 4-38   Backing for Ceiling Material

Offset 2 x 4 studs on 2 x 6 plate help provide
sound barrier

Space between walls

Spaced — double 2 x 4 wall used as sound barrier

Fig. 4-39   Offset Studs –
Soundproofing

partition. The blocking nailed to the partition also serves to brace the wall against lateral loads.

Soundproofing

Many methods of soundproofing have been developed. One which requires special work by the carpenter employs a double row of studs on either a 2 by 6 plate or two 2 by 4 plates set side by side. The studs on each side of the wall are offset to help prevent sound from being transferred to the opposite side of the wall (see Fig. 4-39).

Backing for Vertical Siding

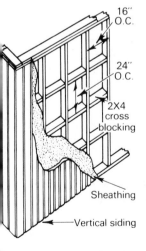

16″ O.C.

24″ O.C.

2X4 cross blocking

Sheathing

Vertical siding

Fig. 4-40   Backing for Vertical Siding

Vertical siding requires some type of wood backing to which the boards can be securely fastened. One method of providing this backing is illustrated in Fig. 4-40. Placing 2 by 4 blocking between the studs 24″ O.C. provides a means for nailing vertical siding securely to the building. These blocks are installed on either side of chalk lines on the studs and fastened with two 16d nails at each end. The one disadvantage to this backing method is the high cost of installation.

Less costly (and less satisfactory) methods of providing backing for vertical siding employ the use of 1 by 4, 1 by 6, or 1 by 8 boards. These boards are placed directly over the studs or the sheathing and are fastened to the studs with 8d or

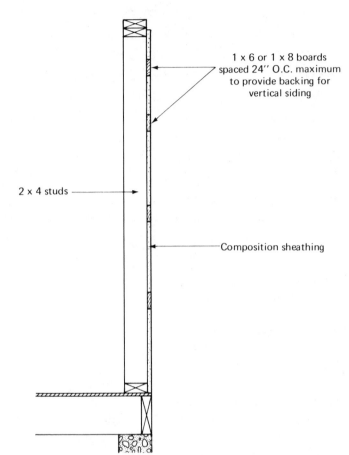

1 x 6 or 1 x 8 boards
spaced 24″ O.C. maximum
to provide backing for
vertical siding

2 x 4 studs

Composition sheathing

Fig. 4-41   Backing for Vertical Siding – Alternate Method

10d nails as required (see Fig. 4-41). The boards placed 24″ O.C. provide satisfactory backing for vertical siding but have less nail-holding power than 2 by 4's placed between the studs.

SCAFFOLDING

A scaffold may be thought of as being a temporary structure used to support workmen, tools and equipment, and materials during the construction of a building. There are many types of scaffolds used by carpenters, and the type of scaffold chosen for a given job will depend on the type of work being done and the size of the job. Small jobs usually require less extensive scaffolding than large jobs, but this is not always true.

SCAFFOLDING SAFETY RULES

as Recommended by

SCAFFOLDING AND SHORING INSTITUTE

(SEE SEPARATE SHORING SAFETY RULES)

Following are some common sense rules designed to promote safety in the use of steel scaffolding. These rules are illustrative and suggestive only, and are intended to deal only with some of the many practices and conditions encountered in the use of scaffolding. The rules do not purport to be all-inclusive or to supplant or replace other additional safety and precautionary measures to cover usual or unusual conditions. They are not intended to conflict with, or supersede, any state or local statute or regulation; reference to such specific provisions should be made by the user. (See Rule II.)

I. **POST THESE SCAFFOLDING SAFETY RULES** in a conspicuous place and be sure that all persons who erect, dismantle or use scaffolding are aware of them.

II. **FOLLOW ALL STATE, LOCAL AND GOVERNMENT CODES, ORDINANCES AND REGULATIONS** pertaining to scaffolding.

III. **INSPECT ALL EQUIPMENT BEFORE USING**—Never use any equipment that is damaged.

IV. **KEEP ALL EQUIPMENT IN GOOD REPAIR.** Avoid using rusted equipment—the strength of rusted equipment is not known.

V. **INSPECT ERECTED SCAFFOLDS REGULARLY** to be sure that they are maintained in safe condition.

VI. **CONSULT YOUR SCAFFOLDING SUPPLIER WHEN IN DOUBT**—scaffolding is his business, **NEVER TAKE CHANCES.**

A. **PROVIDE ADEQUATE SILLS** for scaffold posts and use base plates.

B. **USE ADJUSTING SCREWS** instead of blocking to adjust to uneven grade conditions.

C. **PLUMB AND LEVEL ALL SCAFFOLDS** as the erection proceeds. Do not force braces to fit—level the scaffold until proper fit can be made easily.

D. **FASTEN ALL BRACES SECURELY.**

E. **DO NOT CLIMB CROSS BRACES.**

F. **ON WALL SCAFFOLDS PLACE AND MAINTAIN ANCHORS** securely between structure and scaffold at least every 30' of length and 25' of height.

G. **WHEN SCAFFOLDS ARE TO BE PARTIALLY OR FULLY ENCLOSED,** specific precautions must be taken to assure frequency and adequacy of ties attaching the scaffolding to the building due to increased load conditions resulting from effects of wind and weather. The scaffolding components to which the ties are attached must also be checked for additional loads.

H. **FREE STANDING SCAFFOLD TOWERS MUST BE RESTRAINED FROM TIPPING** by guying or other means.

I. **EQUIP ALL PLANKED OR STAGED AREAS** with proper guard rails, and add toeboards when required.

J. **POWER LINES NEAR SCAFFOLDS** are dangerous—use caution and consult the power service company for advice.

K. **DO NOT USE** ladders or makeshift devices on top of scaffolds to increase the height.

L. **DO NOT OVERLOAD SCAFFOLDS.**

M. **PLANKING:**
1. Use only lumber that is properly inspected and graded as scaffold plank.
2. Planking shall have at least 12" of overlap and extend 6" beyond center of support, or be cleated at both ends to prevent sliding off supports.
3. Do not allow unsupported ends of plank to extend an unsafe distance beyond supports.
4. Secure plank to scaffold when necessary.

N. **FOR ROLLING SCAFFOLD THE FOLLOWING ADDITIONAL RULES APPLY:**
1. **DO NOT RIDE ROLLING SCAFFOLDS.**
2. **REMOVE ALL MATERIAL AND EQUIPMENT** from platform before moving scaffold.
3. **CASTER BRAKES MUST BE APPLIED** at all times when scaffolds are not being moved.
4. **CASTERS WITH PLAIN STEMS** shall be attached to the panel or adjustment screw by pins or other suitable means.
5. **DO NOT ATTEMPT TO MOVE A ROLLING SCAFFOLD WITHOUT SUFFICIENT HELP**—watch out for holes in floor and overhead obstructions.
6. **DO NOT EXTEND ADJUSTING SCREWS ON ROLLING SCAFFOLDS MORE THAN 12".**
7. **USE HORIZONTAL DIAGONAL BRACING** near the bottom and at 20' intervals measured from the rolling surface.
8. **DO NOT USE BRACKETS ON ROLLING SCAFFOLDS** without consideration of overturning effect.
9. **THE WORKING PLATFORM HEIGHT OF A ROLLING SCAFFOLD** must not exceed four times the smallest base dimension unless guyed or otherwise stabilized.

O. For **"PUTLOGS"** and **"TRUSSES"** the following additional rules apply:
1. **DO NOT CANTILEVER OR EXTEND PUTLOGS/TRUSSES** as side brackets without thorough consideration for loads to be applied.
2. **PUTLOGS/TRUSSES SHOULD EXTEND AT LEAST 6"** beyond point of support.
3. **PLACE PROPER BRACING BETWEEN PUTLOGS/TRUSSES** when the span of putlog/truss is more than 12'.

P. **ALL BRACKETS** shall be seated correctly with side brackets parallel to the frames and end brackets at 90 degrees to the frames. Brackets shall not be bent or twisted from normal position.

Q. **ALL SCAFFOLDING ACCESSORIES** shall be used and installed in accordance with the manufacturers recommended procedure. Accessories shall not be altered in the field.

The most simple type of scaffold may employ solidly built wood horses and scaffold planks, while the most complex will employ tubular-steel scaffold frames, braces, railings, and planking. All scaffolds should be built with the safety of the user as a most important consideration.

The set of common sense safety rules on the preceding page has been developed by the Steel Scaffolding and Shoring Institute to promote safety in the use of steel scaffolding. Many of these rules can be applied to all types of scaffolds. Therefore they have been reproduced here to further promote safety in the use of scaffolds.

REVIEW QUESTIONS

1. Why is knowledge of trade terminology important?
2. Describe the two kinds of partitions.
3. What are the functions of a stud?
4. How is stud length determined?
5. What are the characteristics of a corner post?
6. Why is backing needed at wall intersections?
7. How are door and window openings framed?
8. What is the purpose of a wall header?
9. How is the size of rough openings for doors and windows determined?
10. Outline the procedure followed to lay out wall plates.
11. List and describe several types of wind bracing. Which is the best type of brace?
12. What is the purpose of wall sheathing? Name several types.
13. Outline the procedure followed in building a wall.
14. How may walls be aligned and squared?
15. How may walls be erected?
16. How may walls be soundproofed?
17. Why is backing needed for vertical siding?
18. Define a scaffold.

roof

construction

5

The main purpose of the roof is to protect the building against the weather, but the roof may also be an architectural feature which gives the building a desired appearance.

Layout and cutting of rafters requires the use of the steel square and some knowledge of geometry and trade mathematics. Ability to visualize inclined planes and the location of various types of rafters is also necessary (see Fig. 5-1). Carpenters who become proficient in rafter layout are respected by others in the trade and often earn a premium.

TYPES OF ROOFS

There are several types of roofs that the carpenter may be called upon to lay out and build (see Fig. 5-2). The shed roof is a simple design which slopes in only one direction. Gable roofs slope in two directions and are perhaps the most commonly used. All the rafters in the gable roof are identical and are called common rafters.

Fig. 5-1 Kinds of Rafters

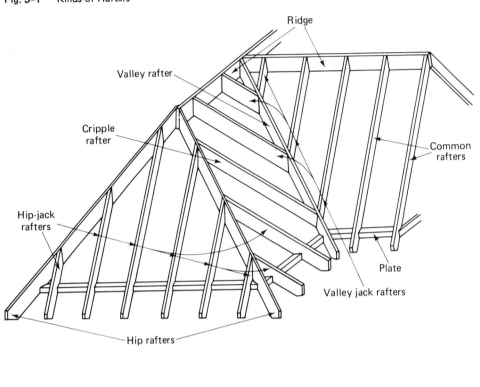

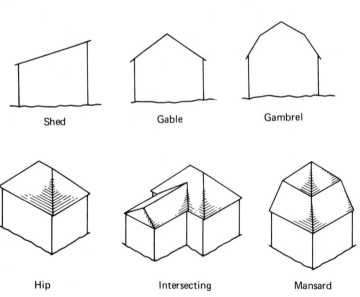

Shed

Gable

Gambrel

Hip

Intersecting

Mansard

Fig. 5-2 Types of Roofs

The hip roof gets its name from the "hips" at each corner of the building. This roof slopes in four directions and, when placed on a rectangular building, will contain some common rafters. A hip roof built on a square structure is sometimes referred to as a pyramid roof.

When it is necessary to gain the greatest possible use of upper-floor space a gambrel roof is sometimes used. This roof has four slopes and is made up of two separate sets of common rafters. Generally the upper set of rafters is comparatively flat while the lower set is quite steep. The manner in which the rafters are fastened will vary with the design and intended use of the building. The gambrel roof is also referred to as a barn roof.

The Mansard roof combines the features of the gambrel and hip roofs and may be built in a number of ways. The lower portion of the roof is quite steep and may be nearly vertical. The upper portion may be either a flat deck, or it may be built with a comparatively small slope.

RAFTER FRAMING PRINCIPLES

Rafter framing is based on the right triangle, and the basic rules for rafter framing are based on the cross section of a gable roof. Figure 5–3 shows the cross section of a gable roof with all parts

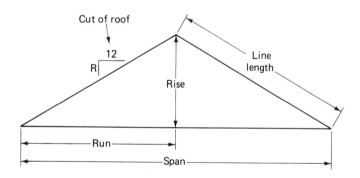

Span — building width
Run — one half span
Rise — distance rafter rises above plate
Line length — mathematical length of rafter (before shortening)

Pitch = $\frac{rise}{span}$

Fig. 5–3   Roof Framing Terms

represented by lines. The following terms are used in discussing roof framing.

Span   The horizontal distance covered by a pair of common rafters. In roof framing span is equal to the width of the building and is one of the first items the roof framer must know before he can lay out the rafters.

Run   The horizontal distance covered by the common rafter. It is equal to one-half the building span. Common rafter run is determined by dividing the building span by 2.

Total Rise   The vertical distance the rafters rise above the wall. It will vary with the angle of roof incline and the width of the building.

Line Length   The mathematical or theoretical length of the rafter. It is the length of the rafter based on a line and does not consider the thickness of rafter stock or framing details.

Cut of Roof   The relationship of unit rise to unit run. It is the slope of the roof and is given on the cross section or elevation as the amount of rise per foot of run.

The shaded area in Fig. 5–4 shows the theoretical triangle on which the length of the rafter is calculated. When the rafter is laid out the line-length side of the theoretical triangle falls inside the rafter stock. Its actual location will vary depending on different roof pitches and cornice designs. It is not necessary to know the actual location of this line, but only to realize that it exists and that it is parallel to the edges of the rafter stock.

When the rafter is laid out the length is marked on the top

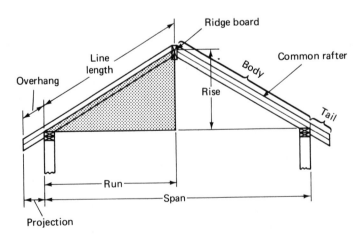

Fig. 5–4   Common Rafter Terminology

edge and a plumb line is drawn on the side of the rafter at each line-length mark. Since the plumb lines are parallel to each other the length of the rafter is the same when measured on the top edge or on any line parallel to the top edge of the rafter stock.

The angle of incline of the roof may be indicated by the slope triangle placed on the elevation or wall cross section, or it may be indicated by a fraction known as the pitch of the roof.

Pitch   The ratio between rise and span. A roof with a total rise of 5′ and a total span of 20′ would have a pitch of $\frac{5}{20}$ or ¼ pitch. The slope triangle gives a vertical distance in relation to a horizontal distance and is often referred to as the cut of the roof.

Rafters can be laid when the total rise and total run are known, but to simplify rafter layout units of rise and run have been established. The relationship between these units is illustrated in Fig. 5–5.

Unit Run   The unit run has been established to be 12″ or 1′. The unit run for any common rafter or jack rafter is always 12″ or 1′. When we were working with total rise and total run, the total run was equal to one-half the span. Following that rule, the unit run is equal to one-half the unit span. Therefore the Unit Span is 2 times 12″ or 24″.

Unit Rise   The unit rise or rise per foot of run is selected to meet the requirements of the roof and the overall design of the building. A small unit rise is used for the comparatively flat roofs on ranch-style homes; large unit rises are used on

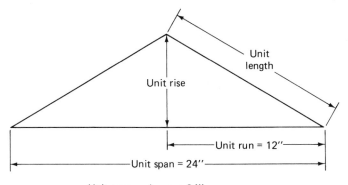

Unit span — always = 24″
Unit run — always = 12″
Unit rise — changes with slope or pitch
Unit length — increases with rise

Fig. 5-5   Roof Framing Units

steeper roofs, making it possible to gain added floor space on the second floor.

Unit Length   The unit length is the line length of the rafter for 1′ of run and 1 unit of rise. The unit length for unit rises from 2″ to 18″ are given on the rafter framing table of the steel square.

When it is necessary to change a fractional pitch to a unit rise the carpenter must remember that pitch is defined as the ratio of rise and span. Therefore pitch is equal to total rise divided by total span, and it is also equal to unit rise over unit span. By setting up a proportion the unit rise can be found as illustrated in the following example.

EXAMPLE

Pitch = ¼;

P = ¼.

Unit span = 24″

$$\frac{\text{Unit rise}}{\text{Unit span}} = \frac{1}{4}$$

$$\frac{\text{Unit rise}}{24''} = \frac{1}{4}$$

Unit rise = 6″

After the carpenter understands the relationship between fractional pitch and unit rise he may take a shortcut to change pitch to unit rise by multiplying the fractional pitch by the unit of span, as illustrated in the following example.

EXAMPLE

Pitch = ¼

Unit span = 24″

¼ × 24″ = ²⁴⁄₄ = 6″ unit rise

Pitch = ⅜

⅜ × 24 = ⁷²⁄₈ = 9″ unit rise

Fig. 5-6   Steel Square – Rafter Framing Table (Courtesy Stanley Tools)

RAFTER TABLES FOR ROOF FRAMING

COMMON RAFTER LAYOUT

Common rafters are usually laid out in two parts. The section over the building is sometimes referred to as the body, and the part which projects outside the building wall is called the tail (see Fig. 5-4).

In common rafter layout the length is calculated for the body of the rafter. This length is marked on the common rafter stock, and additional layout work is done to allow for the thickness of the ridge board, the tail of the rafter, and the seat cut.

When the unit rise is known the line length of the common rafter may be determined by multiplying the unit length found on the rafter framing square by the number of units of run covered by the rafter. The unit length or length of common rafter per foot of run is found on the first line of the rafter table under the inch marks on the outer edge which represent the unit rise (see Fig. 5-6).

EXAMPLE

Span = 26'   Unit rise = 6''

Run = 13'   Unit length = 13.42''

Rafter length = 13.42'' × 13 units = 174.46''

$$0.46'' \times {}^{16}\!/_{16} = \frac{7.36}{16} \text{ or } {}^{7}\!/_{16}$$

174'' ÷ 12'' = 14'6''

Line length = 14'6$^{7}\!/_{16}$''

This line length is transferred to the rafter stock as shown in Fig. 5-7. To allow for the thickness of the ridge board the rafter is shortened. The amount of shortening is equal to one-half the thickness of the ridge board. It is deducted by measuring at right angles to the plumb line. A new plumb line is drawn at the shortened distance. This new plumb line is the cutting line.

The layout for the rafter tail and seat cut is made after the line-length layout is completed. This layout must be made in conjunction with the cornice detail and is discussed in this chapter on pp. 141–143.

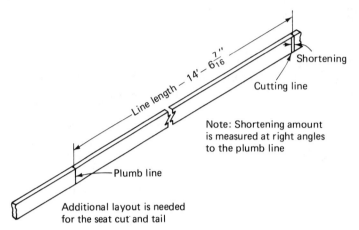

Line length – 14' – 6$\frac{7}{16}$"

Shortening

Cutting line

Note: Shortening amount is measured at right angles to the plumb line

Plumb line

Additional layout is needed for the seat cut and tail

Fig. 5-7   Common Rafter Layout

Some carpenters prefer to lay out the common rafter using the step-off method. In the step-off method the framing square is held with the unit rise along the edge of the rafter on the tongue of the square and the unit run on the body. A mark is made on each side of the square, and the square is stepped off a number of times equal to the run of the rafter. If the rafter has a fraction of a foot of run this fraction is stepped off in the usual manner, but instead of marking at the edge of the rafter for a full unit of run, a mark is placed at the inch mark equal to the fraction of the run (see Fig. 5-8).

While the step-off method of laying out a rafter is easy and theoretically correct it is not as accurate as the unit method because a small amount can be gained or lost on each step. A gain of only $\frac{1}{16}$" on each of the 12 steps will result in a rafter $\frac{3}{4}$" too long. To avoid gaining length on each step a sharp pencil or knife should be used to mark each step, and the square should be aligned carefully for each step.

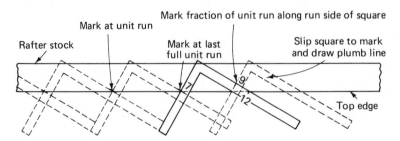

Mark fraction of unit run along run side of square

Mark at unit run

Rafter stock

Mark at last full unit run

Slip square to mark and draw plumb line

9

12

Top edge

Fig. 5-8   Common Rafter Layout – Step-Off Method

LAYOUT OF COMMON RAFTER TAIL

The tail of the rafter is laid out in accordance with the cornice detail. Because this detail is drawn to a small scale it does not give the carpenter all the information he needs. To get the necessary information the carpenter must make a full-size drawing of the cornice showing all its parts and shape of the rafter tail. This drawing is usually made on a sheet of plywood or on the subfloor, but any flat surface can be used.

The length of the common rafter tail can be determined using the unit-length method or the step method. The cornice detail in Fig. 5-9 shows a 24″ projection. The length of the tail can be found by multiplying the unit length by 2 and transferring it to the rafter stock, or by stepping off the unit rise and unit run two times.

After the length of the tail is determined the height of the seat cut and the height of the level cut at the end of the tail must be determined. This information is found by measuring

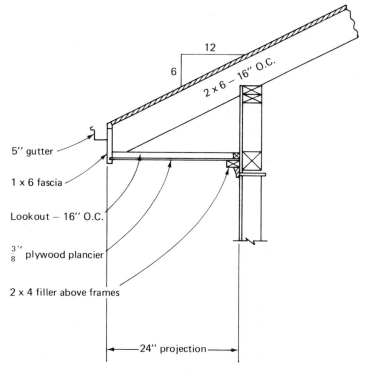

5″ gutter

1 x 6 fascia

Lookout — 16″ O.C.

$\frac{3}{8}$″ plywood plancier

2 x 4 filler above frames

24″ projection

12

6

2 x 6 — 16″ O.C.

Fig. 5-9   Cornice Detail

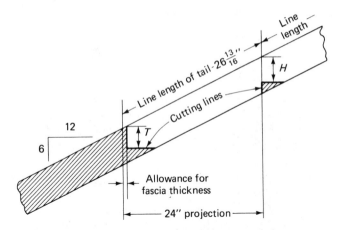

"*H*" is height of seat cut measured along the
plumb line from top of rafter to top of
wall on full size cornice layout and
transferred to rafter pattern

"*T*" is height of tail plumb cut transferred from
the full size cornice detail   Fig. 5-10   Common Rafter – Tail Layout

the height of the seat cut and tail cut on the full-size cornice
drawing and transferring these dimensions to the common rafter
pattern. Figure 5–10 shows the tail of common rafter laid out
to meet the requirements of the cornice in Fig. 5–9. A similar
layout would be made for any size rafter tail by simply transferring dimensions from a full-size drawing of the cornice detail.

It should be noted that in Fig. 5–9 the level cut in the
common rafter seat has full bearing on the top plate. If possible
this level cut should not be longer than the thickness of the wall
to which it is framed.

ROOF PLANS FOR HIP ROOFS AND
INTERSECTING ROOFS

Plain hip roofs and simple intersecting roofs seldom require a
roof plan. However, the beginner in roof framing may find the
roof plan helpful in determining the length of ridge boards and
the locations of the various rafters. The roof plan in Fig. 5–11
shows the location of all the rafters in the roof. Usually the roof
plan will show only the location of the cornice line, wall line,

ridges, hip rafters, and valley rafters, but the student of roof construction may wish to show all the rafters in the roof in order to gain a better understanding of the location and layout of the rafters.

LAYOUT OF RIDGE BOARDS

The ridge board is an aid used in erecting the roof. On it the location of the various rafters is marked. Ridge boards for gable roofs are cut to the same length as the building, but ridge boards for hip roofs must be laid out to make allowances for the various lumber thicknesses and the method used to frame the roof.

The roof in Fig. 5-11 has the hip rafters framed directly to the ridge at one end of the building, but the hip rafters at the other end are framed into three common rafters. Because the hip rafters move into the building at a 45° angle in plan view, the theoretical length of the ridge may be determined by subtracting the width of the building from the length of the building.

This theoretical length is measured on the center line of the ridge board and is the distance between the framing points

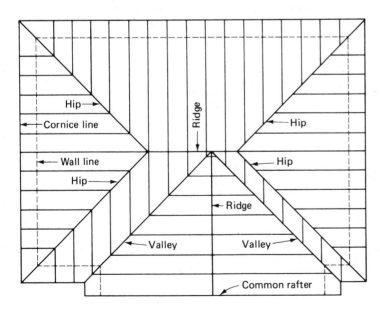

Fig. 5-11   Roof Plan

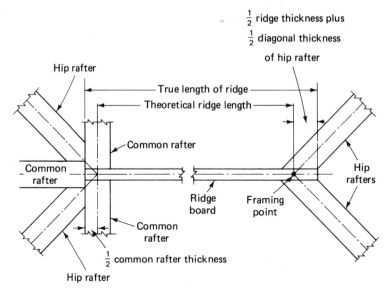

$\frac{1}{2}$ ridge thickness plus

$\frac{1}{2}$ diagonal thickness

of hip rafter

Hip rafter

True length of ridge

Theoretical ridge length

Common rafter

Common rafter

Hip rafters

Ridge board

Framing point

Common rafter

$\frac{1}{2}$ common rafter thickness

Hip rafter

Fig. 5–12   Theoretical and Actual Ridge Length

of the hip rafters. The framing points are created by the inter-section of center lines and are the points to which all mathe-matical or theoretical lengths are calculated.

The actual length of the ridge board is longer than the theoretical length (see Fig. 5–12), and the amount added at each end depends on the method of framing the hip rafters.

When the hip rafters frame directly to the ridge, the ridge must be lengthened by an amount equal to one-half the thick-ness of the ridge plus one-half the diagonal thickness of the hip rafter. The ridge board must be lengthened one-half the thick-ness of the common rafter when hip rafters frame in three com-mon rafters (see Fig. 5–12).

RUN OF THE HIP RAFTER

The hip rafter moves into the building at a 45° angle in plan view. Therefore its actual run is the diagonal of a square with sides equal to the run of the common rafter. It is seldom neces-sary to know the actual run of the hip rafter, because the unit length of the hip rafter which is given on the rafter framing table is calculated for one unit of common rafter run.

Using the common rafter unit run (12″) for the length of the sides of the square of which the diagonal is the hip rafter

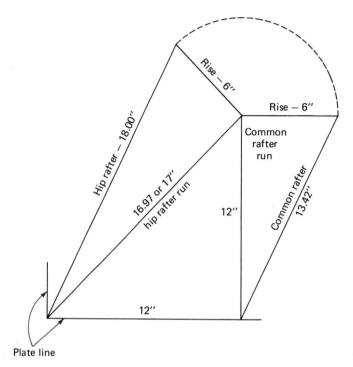

Fig. 5-13   Relationship of Hip Rafter and Common Rafter

unit run, the unit run of the hip rafter may be calculated. Working with two sides of this 12″ square and the diagonal we have a 45° triangle with legs of 12″, and using the Pythagorean theorem we find the length of the third side to be 16.97″. For practical purposes, when doing layout of plumb and level lines 17″ is used as the unit run for all hip rafters and valley rafters in equal-pitch roofs.

Figure 5-13 shows the relationship between the common rafter and hip rafter unit run, unit rise, and unit length. It should be noted that although common and hip rafters have different units of run, they have the same unit rise per unit of run. Because the hip rafter unit run is based on the common rafter unit run hip rafters and common rafters in the same roof have the same number of units of run, rise, and length.

HIP RAFTER LAYOUT

The length of the hip rafter is determined by multiplying the unit length found on the second line of the rafter framing table by the number of units of run covered by the rafter.

Run of common rafter = 13′; unit rise = 4″

Unit length of hip rafter = 17.44″

17.44 × 13 units = 226.72″

$$0.72'' \times \text{^{16}/\textsubscript{16}} = \frac{11.52}{16} \text{ or } \frac{3}{4}''$$

226″ ÷ 12″ = 18′10″

Line length = 18′10¾″

This mathematical line length is transferred to the rafter stock and is laid out on the top edge of the stock in a manner similar to that for common rafters. One line is squared across the stock near the upper end of the rafter and another line is squared across the stock at the line-length mark. Plumb lines are drawn on the side of the rafter stock at the point marked by the lines drawn on the top edge of the stock.

Plumb lines for the hip rafter are laid out by holding the

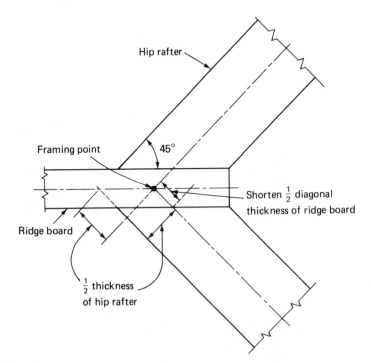

Hip rafter

45°

Framing point

Shorten ½ diagonal thickness of ridge board

Ridge board

½ thickness of hip rafter

Fig. 5-14 Hip Rafters Framing to Ridge – Plan View

unit rise of the roof on the tongue of the square and the unit run of hip rafters, 17″, on the body of the square and marking on the rise side of the square. To lay out level lines, the mark is drawn on the run side of the square.

Bevels or side cuts are laid out on the upper end of the hip rafter; an allowance for the ridge board must be made in accordance with the method by which the hip rafter is framed. When hip rafters are framed directly to the ridge board they meet the ridge as shown in plan view in Fig. 5–14. The line length is figured to the framing point, which is at the center of the ridge board. Therefore the rafter must be shortened an amount equal to one-half the diagonal thickness of the ridge board. Because this is a plan-view distance, it must be deducted by measuring at right angles to the plumb line.

To locate the long point of the bevel a distance equal to one-half the thickness of the rafter stock is marked off as shown in Fig. 5–15. This procedure involves measuring ahead of the line-length plumb line a distance equal to one-half the rafter thickness. This distance is measured at right angles to the plumb line and locates the long point of the rafter before shortening. The amount of shortening is deducted by measuring back at

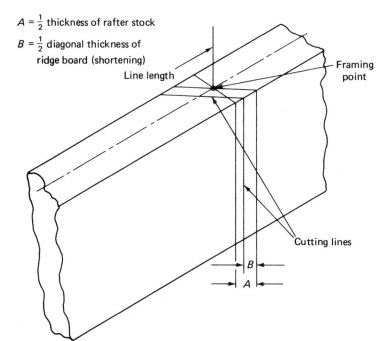

$A = \frac{1}{2}$ thickness of rafter stock

$B = \frac{1}{2}$ diagonal thickness of ridge board (shortening)

Line length

Framing point

Cutting lines

B

A

Fig. 5–15   Layout of Hip Rafter Bevel – Framing to Ridge

right angles from the long point plumb line. The cutting line is thus established.

Some carpenters prefer to shorten the rafter first and then locate the long point plumb line for cutting (see Fig. 5-16). In this procedure the line length is established in the usual manner and marked on the stock. Then the amount of shortening is deducted, measuring back at right angles to the line-length plumb line. A second plumb line is drawn representing the length of the rafter at the center line. To establish the length of the rafter at the long point, it is necessary to measure a distance equal to one-half the rafter thickness ahead at right angles from the shortened plumb line. A third plumb line is drawn at this location. By drawing a line on the top edge of the rafter through the shortened center point to the long point, the bevel is established.

When hip rafters frame into three common rafters as shown in Fig. 5-17, the amount of shortening is equal to one-

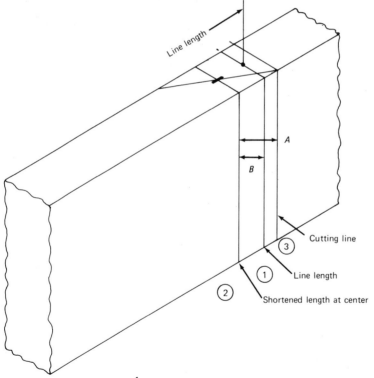

$A = \frac{1}{2}$ Thickness of rafter stock

$B = \frac{1}{2}$ Diagonal thickness of ridge board (shortening)

Fig. 5-16 Layout of Hip Rafter Bevel – Alternate Method

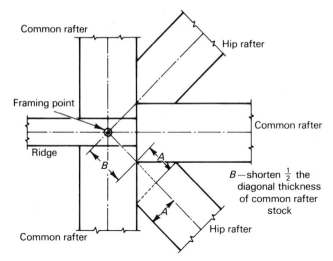

Fig. 5–17 Hip Rafters Framing to Common Rafters – Plan View

half the diagonal thickness of the common rafter. This distance is deducted at right angles to the line-length plumb line and a new plumb line is drawn in. A line squared across the top of the rafter at the new plumb line passes through the long point of the double half-bevel of the rafter. To locate the short point of this half-bevel a distance equal to one-half the rafter thickness is marked off at right angles to the shortening plumb line and a third plumb line is drawn on the side of the rafter. The shape of the bevel may be drawn on the top edge of the rafter stock by connecting the third plumb line with the long point of the double half-bevel located at the center of the rafter on the shortening line (see Fig. 5–18).

The seat cut of the hip rafter must be laid out in a manner which will make its height at the point where the edge of the rafter crosses the plate line equal to the height of the common rafter seat cut. Figure 5-19 shows the plan view of the hip rafter as it crosses the corner of the building. Notice that the line length falls at the corner of the building and that when the line length is squared across the top edge of the rafter it is away from the plate line at the point where it is marked on the side of the rafter.

The distance from the line-length mark to the point where the edge of the rafter crosses the plate line is equal to one-half the rafter thickness. Therefore the height of the hip rafter seat cut must be established on a plumb line drawn at the point where the edge of the rafter crosses the plate line. This new

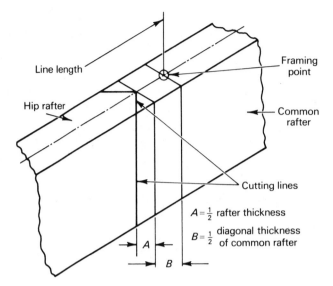

Fig. 5-18   Layout of Hip Rafter Bevels
– Framing to Common Rafters

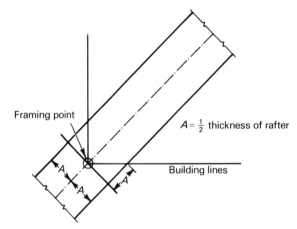

Fig. 5-19   Plan View – Hip Rafter Seat Cut

plumb line is located by measuring in at right angles to the plumb line a distance equal to one-half the rafter thickness. The height of the common rafter seat cut is established on the second plumb line and the level line is drawn in as illustrated in Fig. 5-20.

If the seat cut is not laid out in the manner shown in Fig. 5-20, it is necessary to either drop the rafter or back the top edge of the rafter. Backing is the process of cutting the top edge of the rafter to form a peak at its center. This is done by following a layout line on each side of the rafter with the saw set to come out at the top edge center.

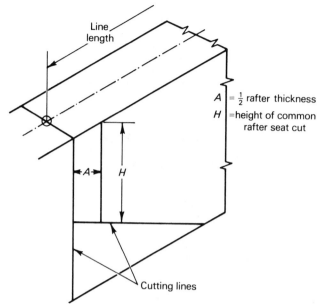

A = $\frac{1}{2}$ rafter thickness

H =height of common rafter seat cut

Fig. 5-20   Layout of Hip Rafter Seat Cut

In backing or dropping the rafter, the height of the common rafter seat cut is marked on the line length mark (see Fig. 5-21). This establishes the height of the rafter at the center line. To establish the height of the rafter at the outside edges, a second plumb line is drawn one-half the rafter thickness in from the line length mark. The height of the rafter at the edge is established by measuring up from the seat cut level line an amount equal to the common rafter seat cut height. The top edge of the rafter may be beveled or backed off. If this is not desirable the rafter is dropped by raising the seat cut level line an amount equal to the backing.

The tail of the hip rafter has the same number of units of run as the common rafter tail in the same roof, but the actual run of the hip rafter tail is equal to the diagonal of the cornice projection and may be determined by multiplying the number of units of cornice projection by 16.97″. To determine the length of the tail the rise and run may be stepped off, or the unit length of the hip rafter may be multiplied by the number of units of cornice projection.

The line length of the tail is marked on the top edge of the hip rafter, starting at the line-length mark at the seat cut as shown in Fig. 5-22. The bevels at the cornice line are laid out so that the long point of the tail falls at the length of the tail mark.

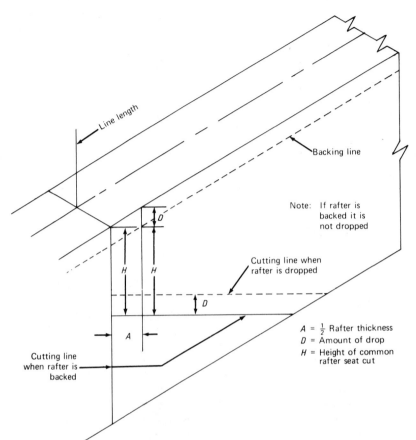

Line length

Backing line

Note: If rafter is backed it is not dropped

Cutting line when rafter is dropped

D

H H

D

A

$A = \frac{1}{2}$ Rafter thickness
D = Amount of drop
H = Height of common rafter seat cut

Cutting line when rafter is backed

Fig. 5-21   Layout for Dropping or Backing Hip Rafter

The horizontal line at the end of the tail is located by transferring the height of the common rafter tail to the cutting lines of the double half-bevel at the tail. Marking at this point results in a hip rafter tail which has the same height as the common rafter tail at the cornice line.

LAYOUT OF HIP JACK RAFTERS

The run of the hip jack is equal to the distance from the corner of the building to the center line of the rafter. If this distance is known, the line length of each rafter may be determined by multiplying the unit length of the common rafter by the jack rafter run. As each jack rafter is a different length this procedure, if carried out for each rafter, could be a lengthy process.

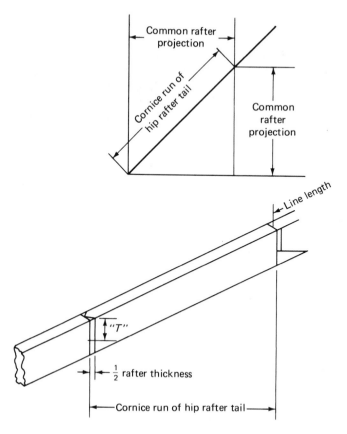

Fig. 5-22   Layout of Hip Rafter Tail

Because jack rafters are uniformly spaced they decrease in length by a constant amount from the longest to the shortest rafter. Therefore it is more desirable to determine the length of the longest jack rafter, mark it on the common rafter pattern, and then mark off the difference in length of jack rafters for the spacing and unit rise being used to determine the length of the remaining rafters.

The difference in length of jack rafters spaced 16″ O.C. is given on the third line of the rafter framing table under the various unit rises on the framing square. It was determined by multiplying the unit length of the common rafter by 1⅓ because 16″ is 1⅓ units of run. The difference in the length of jack rafters spaced 24″ O.C. is found on the fourth line of the rafter table. It was determined by multiplying the common rafter unit length by 2, because 24″ is two units of run. For jack rafters 12″ O.C. the difference in length is the same as the common rafter unit length.

The length of the longest jack rafter may be found by determining where it is placed in the roof in relation to the hip rafter and the ridge board. The simplest way to determine this is to measure on the ridge board the distance from the far side of the common rafter to the location where the long point of the hip rafter bevel will join the ridge. This distance is called C in Fig. 5-23. By subtracting C from the rafter spacing, distance D, which is the amount of run lost by the longest jack rafter, can be determined.

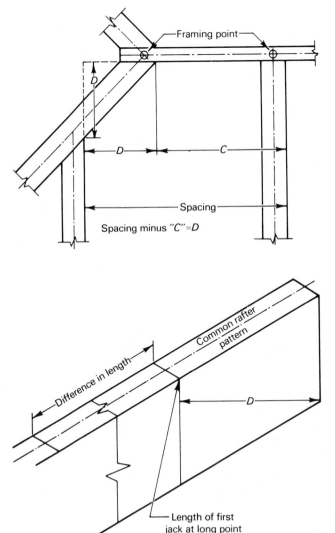

Fig. 5-23   Layout of Hip Jack Rafter Master Pattern

By measuring distance D at right angles to the common rafter pattern· plumb line, the length of the longest jack rafter at the long point is determined, and a plumb line may be drawn at this point to transfer the long point of the jack rafter to the top edge of the rafter stock. It is then squared across the top edge of the rafter pattern, and the common difference in length of jack rafters for the given unit rise and spacing is marked off for the remaining length of the rafter.

It is usually not necessary to lay out the bevels for a jack rafter, as they are automatically cut by setting the power saw table at 45° and following the plumb line. If jack rafters are cut by hand, or if it is necessary to have the shape of the bevel for some other reason, they may be laid out by using the side cuts found on the fifth line of the steel square rafter framing table. When using the side cut table, the number given under the unit rise of the roof is held on one side of the square, 12″ is held on the other side, and the shape of the bevel is marked along the 12″ side.

Bevels or side cuts for jack rafters may also be laid out by measuring back at right angles to the plumb line an amount equal to the rafter thickness and drawing a second plumb line. The length at the second plumb line is squared across the top of the rafter stock, and a diagonal line connecting the two lines squared across the top of the stock represents the shape of the bevel or side cut.

After the master pattern for the jack rafters has been completed on the common rafter pattern, it is necessary to transfer the various lengths to the stock form from which the rafters will be cut. This can easily be done by placing the master pattern alongside rafter stock which has been placed on edge on a set of saw horses. The length is transferred by squaring across the top of the rafter stock at the long point of the bevels and by squaring across the top of the rafter stock at the length mark above the seat cut (see Fig. 5-24).

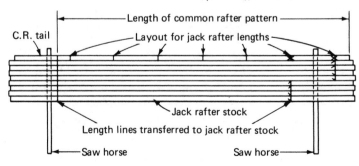

Fig. 5-24   Transfer of Layout Marks from Master Pattern

Following the transfer of the length marks from the master pattern, the plumb lines, seat cut, and tail are laid out with the aid of a tail pattern. The tail pattern is usually made from a board which is the same width as the rafter stock. It is laid out and cut very carefully (see Fig. 5-25). A line squared across the

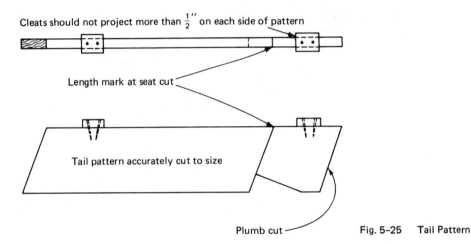

Cleats should not project more than $\frac{1}{2}''$ on each side of pattern

Length mark at seat cut

Tail pattern accurately cut to size

Plumb cut

Fig. 5-25   Tail Pattern

top edge of the tail pattern is used to locate the seat cut at the line-length mark. Small blocks nailed to the top edge of the pattern aid in aligning the top of the pattern with the top edge of the rafter stock and plumb cut at the end of the pattern can be used to draw plumb lines where needed on the jack rafter stock.

Jack Rafter Layout for Hip Rafters
Framing to Common Rafters

When hip rafters frame into the side of common rafters the layout for jack rafters is comparatively simple. This is particularly true when the ridge board is one-half the thickness of the rafters. The length of the longest jack rafter is established by deducting the common difference in jack rafter length from the length of the shortened common rafter pattern. This is possible because the long point of the rafter extended and the far side of the common rafter nearly coincide (see Fig. 5-26). The remainder of the jack rafters are marked off by deducting the common difference in their lengths.

Fig. 5-26   Jack Rafter Layout – Hip Rafters Framed to Common Rafters

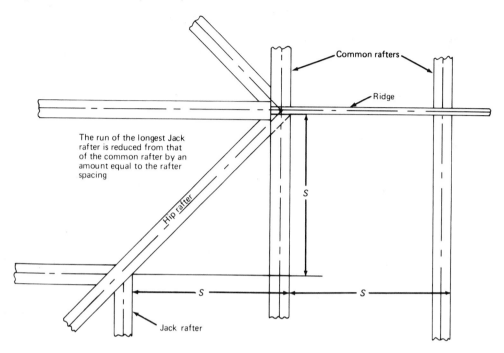

INTERSECTING ROOFS

Buildings which have bays or wings will have intersecting roofs. They may be intersecting hip roofs, intersecting gable roofs, or a combination of the two types. Equal-pitch intersecting roofs fall into two categories. One type of intersecting roof is the same width as the main roof and the other type is narrower than the main roof.

Figure 5-27 shows an enlarged plan view of the valley rafters framing to the ridge boards in an intersecting roof with equal spans. The number of units of run for this valley rafter is the same as the run of the common rafter in this type of roof.

To determine the length of the valley rafter the unit length for the valley rafter is found on the second line of the rafter framing table under the corresponding unit rise on the framing square.

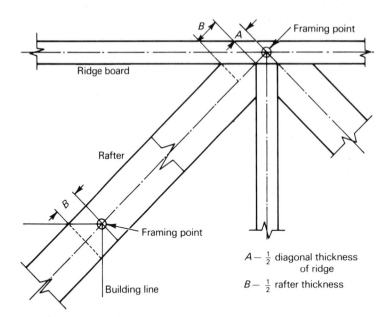

Ridge board

Rafter

Framing point

Building line

$A - \frac{1}{2}$ diagonal thickness of ridge

$B - \frac{1}{2}$ rafter thickness

Fig. 5–27   Valley Rafter Framing – Plan View

EXAMPLE

Unit rise = 8''

Span = 26'

Run = 13'

Unit length of valley rafter = 18.76''

Line length of valley rafter = 18.76 $\times$ 13 = 243.88'' = 20' 3⅞''

The line length of the valley rafter is marked off on the top edge of the valley rafter stock in the same manner as for common rafters and hip rafters, and plumb lines are drawn on the side of the rafter to indicate the line length. Plumb lines for valley rafters are drawn by holding 17'' for the unit run on the body of the square and the unit rise on the tongue of the square and marking along the tongue or rise side of the square. Additional layout at the top end must be made for shortening, to allow for ridge thickness and for the shape of the bevels.

The amount of shortening is equal to one-half the diagonal thickness of the ridge board and is deducted at right angles to the length plumb line at the upper end of the rafter (see Figs. 5-27 and 5-28). To locate the short point of the double half-

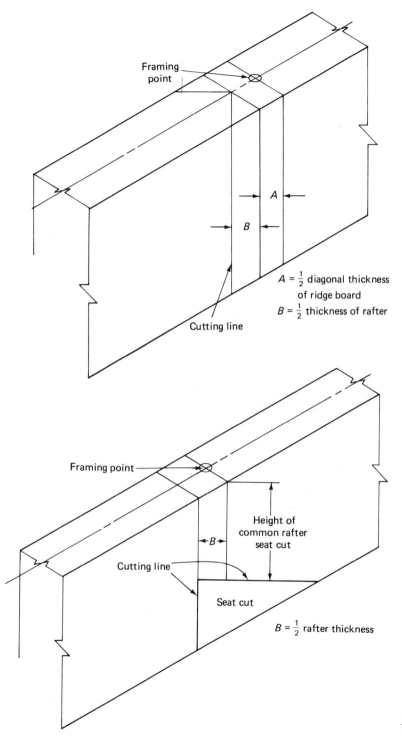

Framing
point

A

B

Cutting line

$A = \frac{1}{2}$ diagonal thickness
of ridge board

$B = \frac{1}{2}$ thickness of rafter

Framing point

Height of
common rafter
seat cut

B

Cutting line

Seat cut

$B = \frac{1}{2}$ rafter thickness

Fig. 5-28   Valley Raf-
ter Layout

159

bevel an additional plumb line is drawn at a distance equal to one-half the rafter thickness from the shortening plumb line.

The line length for the valley rafter is calculated to the framing point which is located at the intersection of the two walls. When the line length is squared across the top of the rafter stock in plan view, it locates the plumb line at the side of the rafter stock and places it on top of the wall. Therefore, when the seat cut is laid out it must be enlarged by an amount equal to one-half the rafter thickness measured at right angles to the plumb line.

The height of the seat cut is established by measuring from the top of the rafter along the length plumb line. The distance from the top edge of the rafter to the level line is equal to the height of the common rafter seat cut (see Fig. 5-28).

The layout for the tail may be accomplished by calculating the length using the run of the cornice and the unit length of the valley rafter. The complete layout of the tail for the valley rafter is illustrated in Fig. 5-28. On many construction jobs the tail is not laid out completely. Instead the tails of the common rafters are allowed to project to the corner of the building and are used to support the roof sheathing.

Intersecting Roofs with Different Spans

Intersecting roofs which have a smaller span than the main roof require a long supporting valley rafter and a short valley rafter. The plan view in Fig. 5-29 illustrates the manner in which these rafters are framed. The length of the long valley is governed by the run of the major span and is marked on the valley rafter stock in the same manner as for other hip or valley rafters.

The amount of shortening for the supporting valley rafter is one-half the diagonal thickness of the ridge board. It is deducted by measuring at right angles to the line-length plumb line. To locate the plumb line at the long point of the bevel an amount equal to one-half the rafter thickness is measured forward at right angles to the shortening plumb line. The plumb line drawn at the long point of the rafter is the cutting line (see Fig. 5-30).

The seat cut for the supporting valley rafter is made in the same manner as the seat cut for valley rafters in intersecting roofs having the same span.

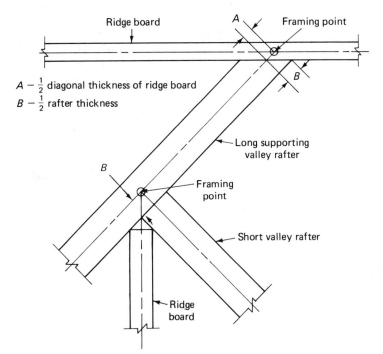

Ridge board

A → Framing point

A − $\frac{1}{2}$ diagonal thickness of ridge board
B − $\frac{1}{2}$ rafter thickness

B

Long supporting
valley rafter

Framing
point

B

Short valley rafter

Ridge
board

Fig. 5-29   Framing Long and
Short Valley Rafters − Plan
View

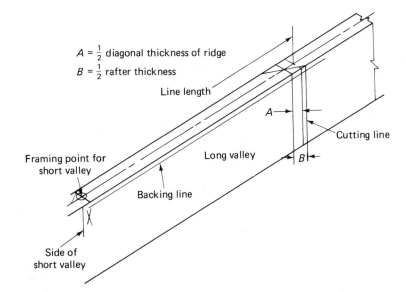

A = $\frac{1}{2}$ diagonal thickness of ridge
B = $\frac{1}{2}$ rafter thickness

Line length

A

Cutting line

Framing point for
short valley

Long valley

Backing line

B

Side of
short valley

Fig. 5-30   Layout of
Long Valley Rafter

Backing the Supporting Valley Rafter   The supporting valley rafter must be backed off from the intersection of the short valley up to the ridge board on the long point edge. This backing is required to avoid a high spot in the roof above the intersection of the two rafters. It is needed because the height of the common rafter is established at the center of the rafter and not at the edges. Therefore, when the rafter meets the ridge the center of the rafter is at ridge height, and the long point edge will be above the ridge unless it is backed off down to the intersection with the short valley.

To determine the amount of backing needed, the framing square is placed on the side of the rafter stock with the unit rise and 17″ held at the edge of the rafter as illustrated in Fig. 5-31. The backing distance is determined by measuring in from the edge of the rafter along the run side of the square a distance of one-half the thickness of the rafter and placing a mark at that point. A line drawn through this point parallel to the edge of the rafter is used as a guide in cutting. The saw is set to follow the line on the side of the rafter and to cut through at the center of the edge of the stock.

Short Valley Rafter   The length of the short valley is governed by the run of the smaller or minor span. It may be stepped off using the unit rise and 17″. The number of steps is

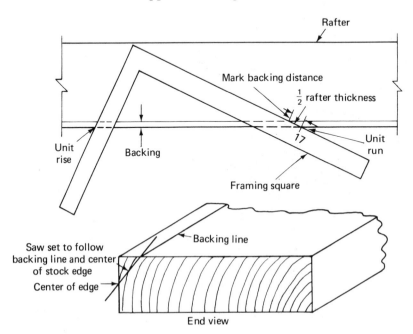

Fig. 5-31   Backing a Rafter

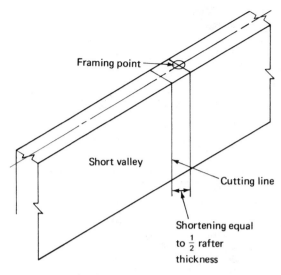

Fig. 5–32   Layout of Short Valley Rafter

(See Fig. 5-25 for seat cut layout)

equal to the number of units of run contained in the minor span.

Using the unit method the length of the short valley is determined by multiplying the unit length of the valley rafter for the corresponding unit rise by the number of units of run contained in the minor span. The line length determined by this method is marked on the rafter stock in the same manner as for other valley rafters. Because the short valley rafter meets the long valley rafter at right angles in plan view, the amount of shortening required is one-half the thickness of the long valley rafter, and the plumb cut does not require a bevel (see Fig. 5–32).

Valley Jack Rafters

The procedure for the layout of valley jack rafters is similar to that for hip jack rafters, the only difference being that the valley jacks do not have a tail, and therefore the lower end is cut at a bevel to fit against the valley rafter.

To determine the length of the longest valley jack it is first necessary to find the distance from the far side of the last common rafter to the side of the valley rafter at the building line. By subtracting this distance from the rafter spacing the amount of run lost by the first jack rafter is determined (see Fig. 5–33).

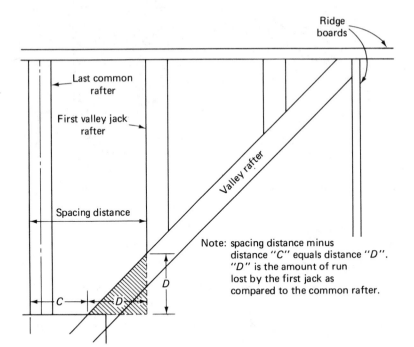

Fig. 5-33   Valley Jack
Rafters – Plan View

The long point of the longest valley jack rafter is determined by deducting distance D at right angles to the line-length plumb line at the seat cut of the common rafter pattern and drawing a new plumb line at that point. The length of the jack rafter is squared across the top of the common rafter pattern, and the difference in length of jack rafters for the corresponding unit rise and spacing is marked off on the top edge to determine the length of the remaining valley jack rafters (see Fig. 5-34).

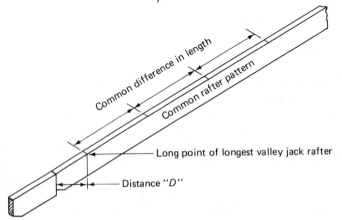

Fig. 5-34   Valley Jack Layout

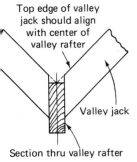

Top edge of valley
jack should align
with center of
valley rafter

Valley jack

Section thru valley rafter

Fig. 5-35   Valley Jacks Fastened to Valley Rafter

When valley jack rafters are nailed in place the top edge of the jack rafter is held even with the top of the ridge board, but it is held above the edge of the valley rafter so that the top of the jack rafter is in line with the center of the valley rafter (see Fig. 5-35).

Cripple Jack Rafters

Cripple jack rafters running from a valley rafter to a hip rafter have a run equal to the distance from the center of the hip rafter to the center of the valley rafter measured along the plate line. The line length may be determined by multiplying the common rafter unit length by the number of units of run covered by the cripple rafter.

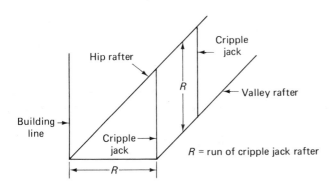

Plan view

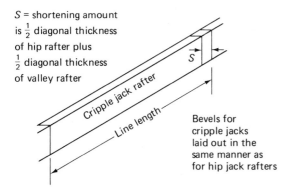

S = shortening amount
is $\frac{1}{2}$ diagonal thickness
of hip rafter plus
$\frac{1}{2}$ diagonal thickness
of valley rafter

Bevels for
cripple jacks
laid out in the
same manner as
for hip jack rafters

Fig. 5-36   Cripple Jack Rafter Layout

Line length may also be determined by use of the step-off method. When the step-off method is used one step of common rafter unit rise and unit run is made for each foot of run covered by the cripple rafter. The layout of a cripple jack rafter is illustrated in Fig. 5–36.

SHED DORMERS

Roofs with shed dormers are comparatively easy to lay out and build. To illustrate the procedure the cross section of the roof in Fig. 5–35 will be used.

The layout of this type of roof begins with the layout of the dormer wall heights. In this problem the main roof has a total rise of 10'6''. This total rise was determined by multiplying the unit rise, 9'', by the number of units of run, which is 14 in this example.

The total rise of the dormer roof is found in a similar manner. The unit rise of 3'' is multiplied by 14 units to get a total rise of 42''. By subtracting the total rise of the dormer roof from the total rise of the main roof the difference in the height of the walls is determined. In this case it is 84'' from the top of the wall on which the main rafters rest to the top of the dormer wall.

The line length of the dormer rafter is found by applying the rules for the layout of a common rafter, and the rafter is shortened in the usual manner. When the seat cut is laid out, the

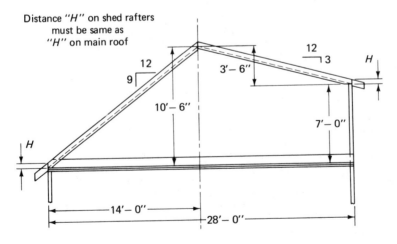

Fig. 5-37   Shed Dormer

height of the seat cut (*H* in Fig. 5–37) measured along the plumb line from the top of the rafter to the level line must be the same as for the height of the seat cut for the common rafters in the main roof.

ORNAMENTAL GABLES

Often a building will have an intersecting roof which is ornamental. This type of roof does not require valley rafters. Instead it is built over the top of the roof sheathing which has been placed over the common rafters. Rafters which are fastened to the roof sheathing require a bevel cut to fit the roof slope along the level line. This is illustrated in Fig. 5–38.

The bevel is laid out by measuring in along the level line, marking a distance equal to the thickness of the rafter stock on the level line, and drawing a plumb line from that point up to the top edge of the rafter stock. A line is squared across the top edge of the rafter stock at the plumb line, and the bevel is obtained by drawing a line diagonally from the level line to the line squared across the rafter. This layout can be applied to any

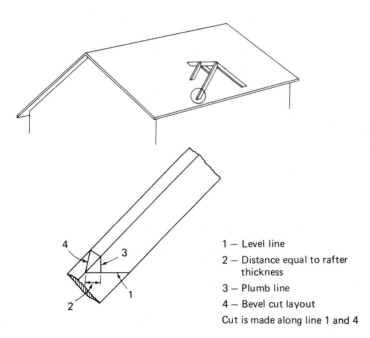

1 — Level line
2 — Distance equal to rafter
　　thickness
3 — Plumb line
4 — Bevel cut layout

Cut is made along line 1 and 4   **Fig. 5–38**   Ornamental Gable

Table 5-1. Angle of incline with the horizontal for common roof slopes

PITCH	CUT	INCLINE IN DEGREES
$\frac{1}{12}$	2 and 12	$9\frac{1}{2}$
$\frac{1}{8}$	3 and 12	14
$\frac{1}{6}$	4 and 12	$18\frac{1}{2}$
$\frac{5}{24}$	5 and 12	$22\frac{1}{2}$
$\frac{1}{4}$	6 and 12	$26\frac{1}{2}$
$\frac{7}{24}$	7 and 12	$30\frac{1}{4}$
$\frac{1}{3}$	8 and 12	$33\frac{3}{4}$
$\frac{3}{8}$	9 and 12	37
$\frac{5}{12}$	10 and 12	40
$\frac{11}{24}$	11 and 12	$42\frac{1}{2}$
$\frac{1}{2}$	12 and 12	45
$\frac{13}{24}$	13 and 12	$47\frac{1}{4}$
$\frac{7}{12}$	14 and 12	$49\frac{1}{2}$
$\frac{5}{8}$	15 and 12	$51\frac{1}{2}$
$\frac{2}{3}$	16 and 12	$53\frac{1}{4}$
$\frac{17}{24}$	17 and 12	$54\frac{3}{4}$
$\frac{3}{4}$	18 and 12	$56\frac{1}{4}$

pitch roof when the main and ornamental roofs have the same pitch.

When the angle of incline with the horizontal is known and the rafter is cut with a portable power saw, the only layout needed is that of the level line. The table of the saw is set to the angle of incline, and the cut is made by following the level line. The angle of incline with the horizontal for various roofs is given in Table 5-1.

RAFTER INSTALLATION

Common rafters are generally fastened in place by toenailing through the seat cut into the top plate and face nailing the rafter to the ceiling joists. On steep roofs and wide buildings, where the ridge board is high above the ceiling joists, it is necessary for the carpenters to build a scaffold at the center of the building to provide a convenient work platform for nailing the rafters to the ridge board. The top edges of the common rafters are held flush (even) with the top edge of the ridge board and fastened with three 8d nails.

Hip rafters and valley rafters are fastened by toenailing at the plate and by nailing to the ridge or common rafters with a sufficient number of 8d and 16d nails.

Hip jack rafters are nailed to the plate in a manner similar to that for common rafters. The upper end of the hip jacks are usually fastened to the hip rafter with three 8d nails. The top edge of the hip jack rafter is kept flush with the top edge of the hip rafter.

Valley jack rafters are fastened to the ridge board in the same manner as common rafters. The lower end of the valley jack rafters are usually nailed to the valley rafter with three 8d nails. The top edge of the valley jack should be aligned with the center of the valley rafter (see Fig. 5-35).

Cripple jack rafters are fastened to the hip rafter in the same manner as hip jack rafters. They are fastened to the valley rafter in the same manner as valley jack rafters.

Various building codes make different requirements in the number and size of nails to be used for the fastening of rafters. Therefore it is necessary for the carpenter to know what the code requirements are in the area in which he is working.

ROOF TRUSSES

Roof trusses offer advantages in speed of erection and building design. However, to use trusses to greatest advantage the roof design is limited to a simple gable.

Trusses may be built at the job site, or they may be purchased from local suppliers and delivered to the job ready for erection. When building roof trusses on the job a level area large enough to accommodate the truss should be provided. Many times the subfloor of the building can be used to build trusses before the walls are erected.

ROOF SHEATHING

Nominal 1″-thick boards 6″ and 8″ wide are commonly used for roof sheathing. These boards are applied at right angles to the rafters, and the end joints between boards occur at the center of a rafter. When placing roof sheathing, four or five consecutive boards may usually be allowed to lap on the same rafter before staggering the joints. Applying boards in this manner speeds up the process of installing roof sheathing.

It is common practice to place roof sheathing with no spaces between the edges of the adjacent boards. When placed in this manner the boards provide good support for all types of roofing material.

Boards 6″ wide are usually nailed with two 8d nails at each rafter, but most building codes require three 8d nails for 8″ boards. The additional nail is needed to prevent warping of the boards.

Plywood roof sheathing will vary in thickness with the spacing of the rafters, the slope of the roof, and the local building codes. Nearly all plywood used for roof sheathing will measure 48″ by 96″. It is installed with the face grain running at right angles to the rafters. All end joints are placed over the center of the rafter.

Plywood sheets ½″ inch thick may be nailed with 6d nails. These nails may be spaced every 6″ along the ends of the sheets and every 10″ at the intermediate rafters. Various types of power nailers are available to speed the installation of all types of roof sheathing.

ROOFING

Roofing material is applied by specialists called roofers in many areas. However, on many occasions carpenters are called upon to install roofing on residential buildings. Materials installed by the carpenter may include asphalt-saturated felt paper, roll roofing, asphalt strip shingles, cedar wood shingles, and hand-split cedar shingles.

Asphalt-saturated felt paper is applied directly over the roof sheathing and acts as an "underlayment" to help prevent leaks caused by water backing up under the shingles. This water may be forced under the shingles by high winds or by ice dams caused by partial thawing and freezing of snow and ice on the roof. Each succeeding row of felt paper should be lapped over the previous row by a minimum of 3″.

Asphalt strip shingles should always be applied in accordance with the manufacturer's instructions provided with shingles. All strip shingles are started with a double first row, which may be made up of a starter roll and a row of shingles or a double course of shingles in which the joints have been offset. The type of starter course used will depend on the type of shingle being installed and the availability of materials.

Asphalt roll roofing is usually applied on garages or temporary buildings, but it is sometimes used as the roofing material on dwellings. Roll roofing usually requires a minimum lap of 2″ between rows. The joint is waterproofed by applying an asphalt roofing cement between the two sheets before fastening the overlapped edge in place with galvanized roofing nails which are driven at 3″ to 4″ intervals.

Cedar wood shingles and hand-split shingles should be installed in accordance with the recommendations of the Red Cedar Shingle Bureau.

All wood shingles should be fastened in place with corrosion-resistant nails. Stainless steel nails are desirable but too expensive in most cases. Therefore steel nails galvanized by the hot-dip process are used extensively to install wood shingles. Nails galvanized by other processes are not acceptable because an acid residue on the nail will cause the shingle to deteriorate around the nail and cause the roof to leak. The nails will also corrode and reduce the service life of the roof.

REVIEW QUESTIONS

1. What are some of the prerequisites for rafter layout?
2. Name and describe the commonly used types of roofs.
3. What is the basis for rafter framing?
4. Define the following items:

Span	Pitch
Run	Unit span
Total rise	Unit run
Line length	Unit rise
Cut of roof	Unit length

5. Find the line length of the common rafters for the following conditions:

SPAN	UNIT RISE
26$'$	5$''$
28$'$	4$''$
27$'$	8$''$

6. What is shortening?

7. How is the seat cut and tail of a common rafter laid out?

8. What is the hip rafter unit run?

9. Find the line length of the hip rafters for the following conditions:

SPAN	UNIT RISE
26$'$	5$''$
28$'$	4$''$
27$'$	8$''$

10. Sketch the layout of bevels for hip rafters.

11. What is a side cut?

12. What is the seat cut for a hip rafter?

13. What is a hip jack rafter?

14. How is the difference in length for hip rafters determined?

15. What is a valley rafter? When is it used?

16. How is the length of a valley rafter determined?

17. Why must the long supporting valley be backed?

18. How are rafters installed?

19. How is roof sheathing applied?

20. What types of roofing may be installed by carpenters?

unequal-pitch
intersecting roofs

6

Intersecting roofs which have the same total rise but different spans are called unequal-pitch roofs. In the unequal-pitch intersecting roof the valley rafters do not run across the building at an angle of 45° in plan view. The angle at which the unequal-pitch valley rafter crosses the building is governed by the run of the two unequal spans and will vary as the spans change.

To aid in identifying the various parts of an unequal-pitch roof the span of the main part of the building is referred to as the major span, and the span of the intersecting roof is called the minor span.

Unequal-pitch intersecting roofs of the usual type may be divided into two categories. The first category is the roof which is placed on a building with no rafter tail projection, and the second is the type which has a cornice projection. The second type, requiring a rafter tail, involves additional layout work. Therefore the unequal-pitch roof without a cornice, which is easier to lay out and build, will be discussed first.

UNEQUAL-PITCH INTERSECTING ROOF
WITHOUT CORNICE

The plan view for an unequal-pitch roof without a cornice is illustrated in Fig. 6-1. Line length for the common rafter in the major span is determined by applying the unit method or the step-off method in the usual manner.

The major span is 24'. Therefore the run is 12', and with a unit rise of 9" the line length of the common rafter of the major span is 15'.

Total rise of the main roof is determined by multiplying the unit rise by the number of units of run covered by the main roof. The total rise of the main roof in Fig. 6-1 is 9'0" and is also the total rise of the minor roof.

Common Rafter in Minor Roof

The length of the common rafter in the minor roof is determined by finding the diagonal distance of the total rise and total run of the minor roof. This distance may be scaled on the rafter framing square by using the 1/12 scale on the back of the square. This is done by allowing 1" for each foot of the rise on one side of the square and for each foot of run on the other side of the square and measuring the diagonal distance.

The step-off method may also be employed to determine the line length of the rafter with the 1/12 scale by using the total

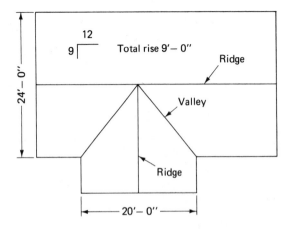

Fig. 6-1   Unequal Pitch Intersecting Roof without Cornice - Plan View

rise on the tongue of the square and the total run on the body of the square and making 12 steps of rise and run to get the line length. The plumb line is established by drawing along the rise side of the square, and level lines are drawn along the run side of the square.

Line length may also be determined by applying the Pythagorean theorem.

EXAMPLE

$$\text{Rafter length} = \sqrt{(\text{total rise})^2 + (\text{total run})^2}$$
$$= \sqrt{(9)^2 + (10)^2} = \sqrt{181}$$
$$= 13'\, 5\frac{7}{16}''$$

When laying out the common rafter of the minor roof the line length is marked on the top edge of the rafter in the usual manner. The rafter is shortened by deducting one-half the ridge thickness at right angles to the plumb line, and the height of the common rafter seat cut is made the same as the height of the seat cut for the common rafters of the main roof (see Fig. 6-2).

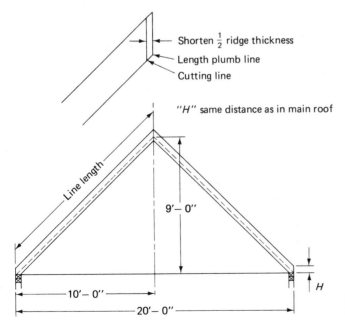

Fig. 6-2   Common Rafter for Minor Roof

Run of Valley Rafter

The run of the valley rafter for this roof is the diagonal of the major run and the minor run. In most cases the actual run of the valley rafter may be scaled using the $\frac{1}{12}$ scale on the framing square, but it can be found mathematically when greater accuracy is necessary.

The valley after run for the roof in Fig. 6-1 is determined in the following example.

EXAMPLE

$$\text{Valley rafter run} = \sqrt{(\text{major run})^2 + (\text{minor run})^2}$$
$$= \sqrt{(12)^2 + (10)^2} = \sqrt{244}$$
$$= 15'7\frac{7}{16}''$$

Length of Valley Rafter

The length of the valley rafter may be determined by finding the diagonal of the total rise and the total run of the valley rafter, either by scaling or mathematically. Figure 6-3 represents the triangle of which the length of the valley rafter is the longest side.

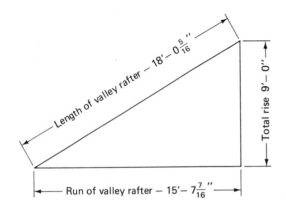

Fig. 6-3   Length of Valley Rafter

$$\text{Length of valley rafter} = \sqrt{(\text{total rise})^2 + (\text{valley rafter run})^2}$$
$$= \sqrt{(9)^2 + (15'7\tfrac{7}{16}'')^2}$$
$$= \sqrt{81 + 244} = \sqrt{325}$$
$$= 18'0\tfrac{5}{16}''$$

As for any other rafter, the line length of the valley rafter is marked on the top edge of the rafter stock and plumb lines are drawn on the side of the rafter at the line-length mark. Plumb lines for the valley rafter are laid out using the $\tfrac{1}{12}$ scale by holding total rise on the tongue of the square and total valley rafter run on the body of the square. If either the total rise or the total run is too big to fit on the square at $\tfrac{1}{12}$ scale, each may be divided by 2 to get numbers which will fit on the square.

Valley Rafter Bevels

After the line length is marked on the valley rafter the bevels must be laid out. To determine the shape of the bevels and the amount of shortening required, a plan view of the two runs is drawn to a scale of $1''$ to a foot (see Fig. 6-4). The valley rafter run is represented by a broken line as are the center lines of the ridge boards. The actual thicknesses of the ridge boards are drawn in with one-half the thickness on each side of the center line. The valley rafter is then drawn in with one-half its actual thickness on each side of the center line representing its run.

A line which runs through the framing point is squared across the top of "rafter" in plan view, and the distances from the length mark to the center of the ridge board are measured and transferred to the rafter stock. These are all plan-view dimensions and are transferred by measuring at right angles to the plumb lines on the rafter stock (see Fig. 6-4).

The shape of the bevel is drawn on the top edge of the rafter by connecting the framing point with the second plumb line, which has established the depth of the bevel on each side of the rafter.

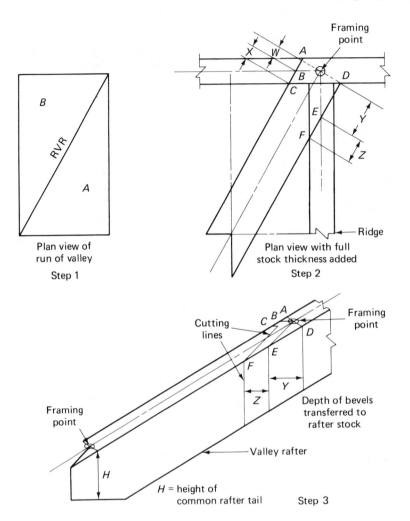

Framing
point

X W A

B D

C

E

F

Y

Z

Ridge

Plan view of
run of valley

Step 1

B

RVR

A

Plan view with full
stock thickness added

Step 2

Cutting
lines

C B A

D

E

F

Z Y

Framing
point

Depth of bevels
transferred to
rafter stock

Valley rafter

Framing
point

H

H = height of
common rafter tail   Step 3

Fig. 6-4   Determining
Bevels for Unequal Pitch
Valley Rafters

Shortening the Valley   Shortening is accomplished by measuring along the edge of the rafter in plan view, from the center
of the ridge to the edge of the ridge. These measurements are
transferred to the rafter stock as shown in Fig. 6-4, and new
plumb lines are marked on the sides of the rafter. Starting at
the final plumb lines, the bevel cutting lines are established by
drawing bevels parallel to those originally laid out on the top of
the rafter. The long point of the bevel at the cutting lines will
not be at the center of the rafter but will be slightly off to one
side. The amount that the long point moves off center will vary
with the various spans of the unequal-pitch intersecting roof.

Jack rafters for each section of the unequal-pitch roof will have a different plumb cut, a different bevel cut, and an uncommon difference in length of jack rafters. The plumb cut is governed by the plumb cut of the common rafters in the section of the roof in which the jack rafter will be installed. Jack rafters in the major span have the same plumb cut as common rafters in the major span, and jack rafters in the minor span have the same plumb cut as common rafters in the minor span.

The bevel cut and the difference in length of jack rafters can best be determined by working with what is called the slant triangle. The slant triangle is a portion of the roof which is bounded by the valley rafter, a portion of the ridge board, and the common rafter which runs from the valley rafter to the ridge board (see Fig. 6-5).

Working with slant triangle *A,* which lies on the main roof, the first step in determining the difference in the length of jack rafters is to find the number of spaces between the rafters included in the ridge length. The length of the ridge in question is that portion cut off by the minor span (see Fig. 6-6). With jack rafters 16″ on center the number of spaces in the ridge would be 120″/16″ or 7½ spaces.

To determine the difference in the length of jack rafters the number of spaces in the ridge is divided into the length of the common rafter in the span under consideration. In Fig. 6-6 the common rafter is 15′0″ or 180″ long. The difference in

LCR — length common rafter
RVR —run valley rafter

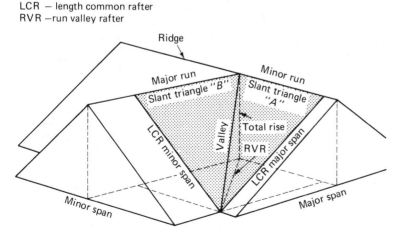

Fig. 6-5   Slant Triangles

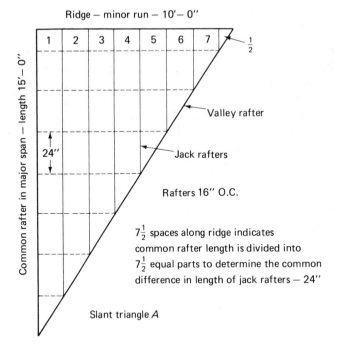

Ridge — minor run — 10′— 0″

Common rafter in major span — length 15′— 0″

24″

Valley rafter

Jack rafters

Rafters 16″ O.C.

$7\frac{1}{2}$ spaces along ridge indicates
common rafter length is divided into
$7\frac{1}{2}$ equal parts to determine the common
difference in length of jack rafters — 24″

Slant triangle *A*

Fig. 6-6   Finding Difference in Length
of Jack Rafters

length of jack rafters, therefore, is 180″/7.5 spaces or 24″.
Notice how the lines projected from the intersection of the jack
and valley rafters divide the common rafter into seven and one-
half parts. In effect this drawing proves that this method for
determining the difference in length of jack rafters is logical.

The same procedure is applied to slant triangle *B* to determine the difference in length of jack rafters for that section
of roof. In this triangle the length of the ridge is equal to the
major run and the length of the common rafter is taken from
the minor span. With the rafters 16″ on center the ridge length,
144″, divided by 16″ indicates that there are nine spaces contained in the ridge.

The difference in the length of jack rafters is found by dividing the length of the common rafter in the minor span by the
number of spaces in the ridge length equal to the major run.
In triangle *B* the common difference in jack rafter length is
161⁷⁄₁₆″/9 spaces or 17¹⁵⁄₁₆″. Notice how the lines projected
from the intersection of the valley rafter and jack rafters in
Fig. 6-7 divide the common rafter of the minor span into nine
equal parts, each 17¹⁵⁄₁₆″ long.

Jack Rafter Bevels   Bevels on the jack rafters may be laid

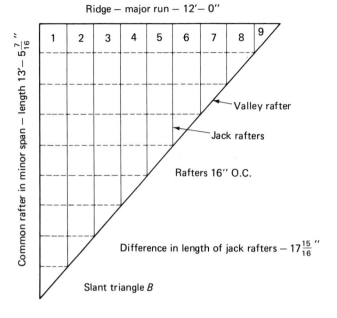

Ridge — major run — 12′– 0″

Common rafter in minor span — length 13′– $5\frac{7}{16}$″

| 1 | 2 | 3 | 4 | 5 | 6 | 7 | 8 | 9 |

Valley rafter

Jack rafters

Rafters 16″ O.C.

Difference in length of jack rafters — $17\frac{15}{16}$″

Slant triangle *B*

Fig. 6-7　Finding Difference in Length of Jack Rafters

out by employing the slant triangles and the $\frac{1}{12}$ scale. The bevel on the jack rafter is governed by the length of the common rafter and the length of ridge into which it frames. For jack rafters in the major span the bevel is controlled by the length of the common rafter in the major span and the length of ridge cut off by the minor run. Therefore the bevel of jack rafters in triangle *A* in Fig. 6-6 is governed by 15 and 10.

To lay out this bevel on the top edge of the jack rafter 10″ is held on the tongue of the square and 15″ is held on the body of the square. The bevel is always marked on the side of the square which represents the rafter length, 15″ in this example (see Fig. 6-8).

The same procedure is applied for the jack rafters in triangle *B* except the length of the ridge is 12′, and the common rafter length is 13′5⅞₆″. Therefore 12″ is held on the tongue of the square, and 13 and 5½ twelfths inches is held on the body of the square. The bevel is marked along the rafter length side of the square.

Shortening is required when jack rafter lengths are calculated to the center lines of the ridge and valley rafter. The amount of shortening needed at the upper end is the same as that required for a common rafter and is laid out in the same manner. The amount of shortening at the lower end can be

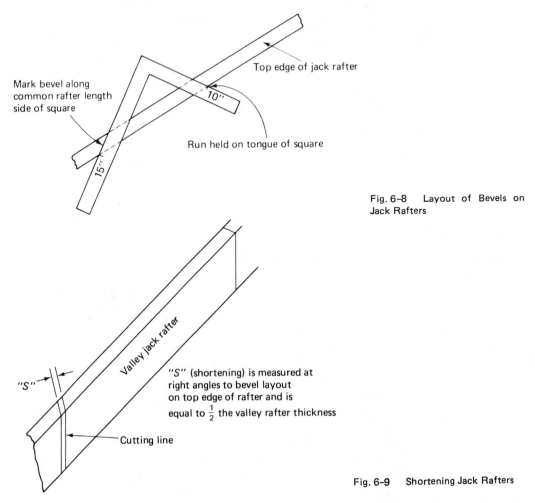

Mark bevel along common rafter length side of square

Top edge of jack rafter

10″

15″

Run held on tongue of square

Fig. 6-8   Layout of Bevels on Jack Rafters

Valley jack rafter

"S"

"S" (shortening) is measured at right angles to bevel layout on top edge of rafter and is equal to $\frac{1}{2}$ the valley rafter thickness

Cutting line

Fig. 6-9   Shortening Jack Rafters

found quickly by measuring at right angles to the bevel on the top edge of the rafter stock and marking off a distance equal to one-half the valley rafter thickness. A new bevel is drawn at this point, and a new plumb line is also marked on the side of the rafter (see Fig. 6–9).

UNEQUAL-PITCH INTERSECTING ROOF WITH CORNICE

The task of laying out and framing an unequal-pitch intersecting roof with a cornice introduces problems not encountered in the

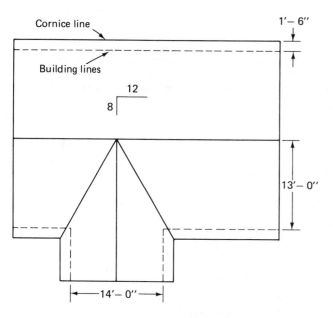

Fig. 6-10   Unequal Pitch Intersecting Roof with Cornice – Plan View

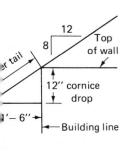

Fig. 6-11   Cornice Drop

roof without a cornice. Generally there is a desire to maintain an equal width and height in the cornice on both sections of the building, but because one roof is steeper than the other it will drop farther below the top of the wall than the rafter tail on the flatter roof. To avoid the problem of unequal height in the cornice, it is necessary to work from the cornice line when determining the lengths of rafters in the minor roof and to raise the top of the wall for the roof in the minor span.

Figure 6-10 gives the plan view for an unequal-pitch intersecting roof with a cornice projection. The length of the common rafter in the main roof and the length of the rafter tail are determined in the usual manner. Before the length of the common rafter in the minor roof can be determined the cornice drop and total increased rise must be known.

The cornice drop is determined by multiplying the unit rise of the main roof by the run of the cornice. For the roof in Fig. 6-10 the cornice drop is 1.5 times 8″, or 12″ (see Fig. 6-11).

To determine the total increased rise the cornice drop is added to the total rise of the main roof. The total increased rise is used when determining the length of the common rafter in the minor roof.

Common Rafter

The total increased run is determined by adding the run of the cornice to the run of the minor span. Total increased run is used when calculating the length of the common rafters in the minor roof.

EXAMPLE

(See Fig. 6–10 and 6–12)

Total rise of main roof = 8'8"

Cornice drop = 1'

Total increased rise = 9'8"

Total increased run of the minor roof = 8'6"

Length of common rafter in minor roof =
$\sqrt{(8.5)^2 + (9.67)^2} = 12'10\frac{7}{16}''$

It should be noted from the comparative drawing in Fig. 6-12 that the common rafter running from the cornice line to the ridge does not meet the top of the normal wall height. Therefore the plate of the intersecting wall must be raised. The amount which the plate must be raised can be determined by making a comparative-pitch drawing and measuring the distance between the two rafter slopes (see Fig. 6-13), but it may also be

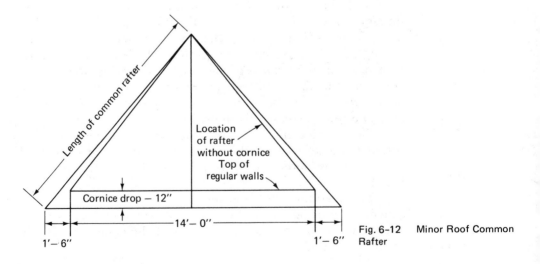

Location of rafter without cornice
Top of regular walls

Length of common rafter

Cornice drop — 12"

1'– 6"

14'– 0"

1'– 6"

Fig. 6-12   Minor Roof Common Rafter

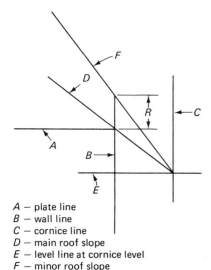

A — plate line
B — wall line
C — cornice line
D — main roof slope
E — level line at cornice level
F — minor roof slope
R — amount of raised plate

Fig. 6-13   Comparative Pitch

determined mathematically by finding the difference between the cornice rise in the minor and major roofs.

To find the height of the raised plate the unit rise of the minor roof must be determined. This is done by dividing the total increased rise of the minor roof in inches by the total increased run in feet.

EXAMPLE

$$\text{Unit rise of minor roof} = \frac{\text{total increased rise (in inches)}}{\text{total increased run (in feet)}}$$

$$= \frac{116}{8.5}$$

$$= 13.65''$$

Rise of minor roof over cornice = 13.65″ × 1.5 = 20.5″

Rise of major roof over cornice = 8″ × 1.5 = 12″

Height of raised plate = 20.5″ - 12″ = 8.5″

The seat cut in the common rafter is located by measuring in from the end of the tail or stepping off the length of the tail from the overall rafter length. When locating the seat cut care must be taken to avoid errors which will cause difficulties in

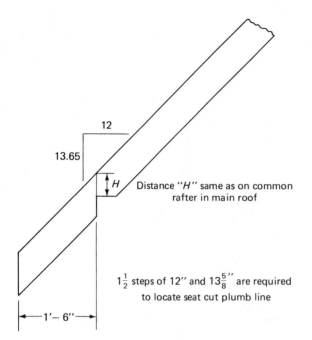

13.65

12

H   Distance "H" same as on common
rafter in main roof

$1\frac{1}{2}$ steps of 12″ and $13\frac{5}{8}$″ are required
to locate seat cut plumb line

1′– 6″

Fig. 6-14   Locating Common Rafter Seat Cut

framing the roof. The layout for locating the seat cut is illustrated in Fig. 6–14. Care must be taken to maintain the proper height of the seat cut if the rafter is to fit properly, and the height of the plumb cut at the tail must be the same as those on the common rafters of the main roof.

Valley Rafter

The length of the valley rafter is based on the diagonal of the total increased runs of the roof and the total increased rise. To find the run of the valley rafter the diagonal of the total increased runs may be scaled, or it may be calculated mathematically. The following example is based on the roof in Fig. 6–10.

EXAMPLE

Run of valley rafter =
$$\sqrt{(\text{increased major run})^2 + (\text{increased minor run})^2}$$
$$= \sqrt{(14.5)^2 + (8.5)^2}$$

$$= \sqrt{282.50} = 16'9^{11}\!/_{16}\,''$$

Length of valley rafter =
$$\sqrt{(\text{increased rise})^2 + (\text{run of valley rafter})^2}$$

$$= \sqrt{(9.67)^2 + 282.50} = 19'4^{11}\!/_{16}\,''$$

Layout of the bevels and shortening for this type of valley rafter is made in the same manner as for unequal-pitch roofs without a cornice. The layout of the seat cut and tail of this rafter requires the use of a drawing of the plan view of the rafter tail made to full scale.

This drawing shows the building lines and cornice lines at the intersecting roof. The center line of the valley rafter is drawn in from the intersection of the cornice lines over to the plate line. The angle at which the center line is drawn is gov-

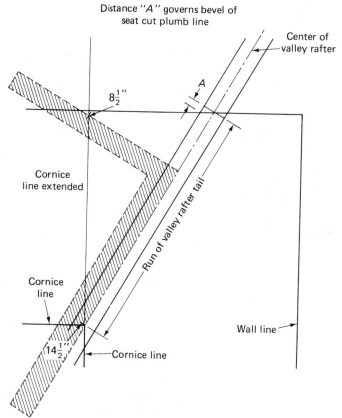

Fig. 6-15   Locating Valley Rafter
Seat Cut

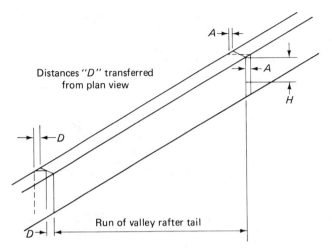

Distances "D" transferred
from plan view

Run of valley rafter tail

Fig. 6-16   Layout of Valley Rafter
Seat Cut

erned by the total increased runs. One run is held on the tongue of the square and the other run is held on the body of the square along the cornice line extended (see Fig. 6–15). The run of the valley rafter tail is measured along the center line in the plan view and is transferred to the valley rafter by stepping off (measuring) at right angles to the plumb line. A plumb line drawn at the distance stepped off represents the length of the rafter tail at its center along the plate line.

Because the rafter meets the wall at an angle the seat cut must be laid out accordingly. The shape of the cut may be taken from the plan view of the valley rafter and transferred to the rafter stock. All plan-view dimensions must be transferred by measuring at right angles to the plumb line (see Fig. 6–16).

Jack rafters for the unequal-pitch intersecting roof with a cornice are laid out in the same manner as jack rafters for the unequal-pitch intersecting roof without a cornice.

UNEQUAL-PITCH INTERSECTING ROOFS WITH
LESS TOTAL RISE THAN MAIN ROOF

Occasionally the carpenter will be called upon to build an unequal-pitch intersecting roof which has less total rise than the main roof. The plan view for this type of roof is illustrated in Fig. 6–17. Although unit rise is given for both sections of the roof, there are problems involved in determining cornice width,

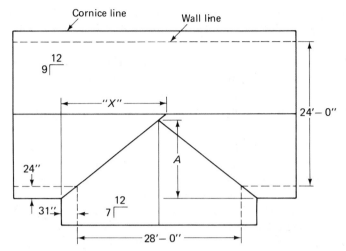

Fig. 6-17   Wide Intersecting Roof

cornice drop, and plate height. Special problems arise in determining the run, rise, and length of the valley rafters.

Twelve steps which may be employed to solve this type of roof in a logical manner are listed and explained in the following discussion.

1. Find cornice drop of the main roof. The cornice drop is determined by multiplying the cornice projection by the unit rise of the main roof.

2. Find cornice drop of the minor roof. The cornice drop of the minor roof is determined by multiplying the cornice projection of the minor roof by the unit rise of the minor roof. Often the cornice width of the minor roof may be adjusted to provide a cornice drop equal or nearly equal to the cornice drop of the main roof.

3. Find the height of the raised or lowered plate. The difference in plate height is found by finding the difference between the two cornice drops. If the cornice drop of the minor roof is greater than that of the major roof the plate of the minor roof must be raised, but if it is less than that of the major roof it must be lowered. If the difference between the plate height is small the plates may be left at the same height, and the difference in cornice drops made up in the height of the seat cuts of the common rafters.

4. Find the length of the common rafter in the major span. This can be done in the usual manner, but to

make calculating the lengths of the valley rafters easier the total length of the rafter including the tail should be determined.

5. Find the length of the common rafter in the minor span. Once again, this can be done in the usual manner, but to facilitate calculating valley rafter lengths the total length of the rafter including the tail should be determined.

6. Find the length of ridge A in the minor span (see Fig. 6-17). The height of the minor ridge must be determined first. After the total height of the minor roof, in inches, is known it is divided by the unit rise of the major roof. The result is the number of units of rise required in the main roof to reach the height of the minor ridge. There is the same number of units of run included in the distance labeled A. Therefore, to determine distance A the total rise of the minor roof is divided by the unit rise of the major roof.

7. Find the length of the short valley rafter. The length of the short valley rafter may be determined by using the slant triangle method. The short valley rafter is the hypotenuse of the triangle which has distance A for one leg and the overall length of the minor common rafter for the other leg. The rafter length may be scaled, or it may be determined mathematically.

8. Find the length of the long valley rafter. The sides of similar triangles are proportional. Therefore the length of the long valley rafter is proportional to the length of short valley rafter, and distance A is proportional to the distance covered by the total increased run of the major roof. A proportion for determining the length of the long valley rafter could be set up as follows:

$$\frac{\text{Length of long valley rafter (LVR)}}{\text{Length of short valley rafter (SVR)}} = \frac{\text{major increased run (MaR)}}{A}$$

$$\text{LVR} = \frac{(\text{SVR}) (\text{MaR})}{A}$$

9. Find the run of the short valley rafter. The run of the short valley rafter is the diagonal of distance A and the total increased run of the minor roof. It may be determined by scaling, or it may be determined mathematically.

10. Find the total rise of the short valley rafter. The rise of the short valley rafter is the same as the total increased rise of the minor roof. It is determined by multiplying the unit rise of the minor roof by the total increased run of the minor roof.

11. Find the cut of the valley rafters. The cut of the valley rafters is determined by the numbers held on the framing square when drawing the plumb lines and level lines. It is governed by the total rise and total run of the valley rafter or a proportional part of the total rise and total run.

12. Find the run X of the long valley. The run X of the long valley is the distance measured along the ridge from the framing point at the cornice intersection to the point where the center line of the valley meets the center line of the ridge. Run X is proportional to the total increased run of the minor roof. Therefore the following proportion may be set up to determine the X distance:

$$\frac{X}{\text{Increased minor run (IMiR)}} = \frac{\text{Increased major run (IMaR)}}{A}$$

$$X = \frac{(\text{IMiR})(\text{IMaR})}{A}$$

To illustrate the use of the preceding 12 steps of solving the unequal-pitch roof which has less total rise than the main roof, the following example based on the plan view in Fig. 6–17 is given.

EXAMPLE

1. Find the cornice drop: main roof: 2 times $9'' = 18''$

2. Find the cornice drop: minor roof: $2\frac{7}{12}$ times $7'' = 18\frac{1}{12}''$

3. Find the height of raised or lowered plate: $18\frac{1}{12}'' - 18'' = \frac{1}{12}''$ raised

   Note: Because of the small difference between the two cornice drops the difference will be made up in the height of the seat cut.

4. Find the length of the common rafter, major span (including tail): $14 \times 15'' = 17'6''$

5. Find the length of the common rafter, minor span (including tail); $16.58 \times 13.89'' = 19'2\frac{5}{16}''$

6. Find the length of ridge A:

$$\frac{16\frac{7}{12} \times 7''}{9} = 12.9 \text{ units} \quad \text{or} \quad 12'10\frac{13}{16}''$$

7. Find the length of the short valley rafter:

$$\sqrt{(12.9)^2 + (19.19)^2} = 23'1\frac{1}{2}'$$

8. Find the length of the long valley rafter:

$$LVR = \frac{(SVR)\,(MaR)}{A}$$

$$= \frac{(23.12)\,(14)}{12.9} = 25'1\frac{1}{8}''$$

9. Find the run of the short valley:

$$RSV = \sqrt{(12.9)^2 + (16.58)^2} = 21'0\frac{1}{8}''$$

10. Find the rise of the short valley rafter:

$$16\frac{7}{12} \times 7'' = 116\frac{1}{12}'' = 9'8\frac{1}{12}''$$

11. Find the cut of the valley rafters:

From 9 and 10 above, rise = $9\frac{8}{12}''$ and run = $21''$
Plumb lines are marked on $9\frac{8}{12}''$
Level lines are marked on $21''$

12. Find run X of the long valley rafter:

$$X = \frac{(IMiR)\,(IMaR)}{A}$$

$$= \frac{(16.58)\,(14)}{12.9} = 18'0''$$

To lay out the bevels for the valley rafter a plan-view drawing of the runs involved is made to a scale of $1''$ per foot. The thickness of the ridges and valleys are drawn in full size with one-half the thickness on each side of the center line in the plan-view drawing. This drawing is illustrated in Fig. 6–18. All

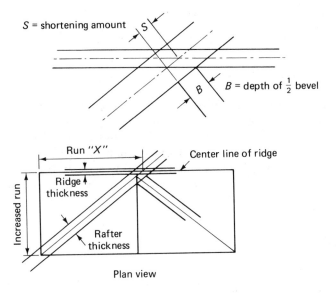

S = shortening amount

B = depth of $\frac{1}{2}$ bevel

Run "X"

Center line of ridge

Ridge thickness

Increased run

Rafter thickness

Plan view

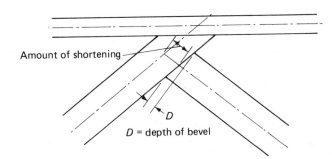

Amount of shortening

D

D = depth of bevel

Fig. 6-18　　Layout of Rafters for Wide Intersecting Roof

dimensions transferred from plan view to the rafter stock must be transferred by measuring at right angles to the rafter plumb line.

The difference in the length of jack rafters in the major span is determined by first dividing distance X by the rafter spacing ($12''$, $16''$, or $24''$) to determine the number of spaces. The number of spaces is divided into the overall length of the common rafter to get the difference in the length of jack rafters.

To find the difference in the length of jack rafters in the minor span the number of spaces in distance A is divided into the overall length of the minor span common rafter.

The bevel on the jack rafters in the main roof is governed by the length of the common rafter of the main roof and distance *X*. The bevel is marked on the common rafter length side of the square.

In the minor roof the bevel on the jack rafters is governed by the length of the common rafter in the minor roof and distance *A*. The bevel is marked on the common rafter length side of the square.

The plumb lines on jack rafters in the major roof are the same as those for the common rafter in the major roof, and the plumb lines on the jack rafters in the minor roof are the same as the plumb lines on the common rafters in the minor roof.

Shortening of the jack rafters is accomplished in the same manner as for jack rafters in other types of unequal-pitch roofs.

REVIEW QUESTIONS

1.   What is an unequal-pitch intersecting roof?

2.   How may the length of the common rafter in the minor span be determined?

3.   How is the run of the valley rafter determined?

4.   How is the length of valley rafters for unequal-pitch roofs determined?

5.   How is the shape of bevels for unequal-pitch rafters determined?

6.   How is the valley rafter shortened?

7.   How is the difference in length of jack rafters determined?

8.   How are bevels for jack rafters in unequal-pitch roofs laid out?

9.   How does layout for the unequal-pitch roof with a cornice differ from layout for the roof without a cornice?

10.   Outline the procedure for laying out an unequal-pitch roof when the intersecting roof has less total rise than the main roof.

exterior trim and exterior siding

7

Exterior trim and exterior siding serve two purposes. They dress up the building and give it its finished appearance, and they also serve to protect the building framework against the weather.

Materials used for exterior finish work should be resistant to the effects of rain, snow, sun, and wind, and should have the ability to accept and hold a finish. Some lumber species which have been found to be resistant to the effects of the weather are redwood, cedar, and various western pines. Various types of plywood, composition board, aluminum, and plastic sidings are also in general use, and the carpenter is called upon to install these as well as the traditional wood materials.

Nails used to fasten materials which are exposed to the weather should be weather resistant and of sufficient size to hold securely. Hot-dipped galvanized nails, aluminum nails, and stainless steel nails are suitable for fastening wood exterior trim and siding. Aluminum or stainless steel nails must be used on aluminum siding to avoid galvanic action where the nail and siding come in contact.

TYPES OF CORNICES

Cornice design may vary in size, shape, and number of component parts, but regardless of design variations cornices may be divided into two categories: the closed or box cornice, and the open cornice.

The cross section of a typical box cornice is illustrated in Fig. 7–1. This type of cornice gets its name from the fact that it encloses or boxes in the rafter tails. The false fascia nailed to the rafter tails serves as backing for the plancier material. It is not needed when the plancier is narrow and made of material which is thick enough to support itself.

The fascia board placed over the false fascia serves as a support for the rain gutter and also provides a finished appearance at the end of the rafter tails.

The plancier runs from the building to the fascia and is sometimes referred to as a soffit. It may be made from many different materials.

Lookouts are used to support the plancier. They are usually 2 by 2 or 2 by 4 members which run from the building to the fascia. They may be spaced from 16″ to 48″ on center. The spacing of the lookouts varies with the thickness and width of the plancier material.

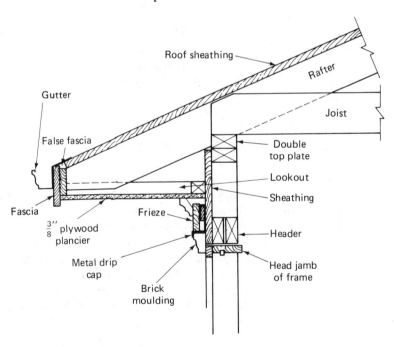

Fig. 7–1   Typical Box Cornice

The frieze board fills in the space between the top of the window frames and the soffit. The width of the frieze and the number of members, moldings, and so on which are a part of the frieze will vary with the cornice design.

A molding is always used to cover the joint between the frieze and the plancier.

A closed cornice with a sloping plancier (Fig. 7–2) has the plancier fastened to the underside of the rafter tails, thereby eliminating the need for lookouts. Some difficulty is encountered in fitting the joint between the plancier and fascia.

Another type of closed cornice often called a snub cornice is illustrated in Fig. 7–3. This is the most economical type of cornice to build since it contains only a fascia board fastened to the end of the rafters.

The open cornice utilizes exposed rafter tails and exposed roof sheathing (see Fig. 7–4). The rafter tails may be specially made pieces called outlookers which are nailed to the side of the rafter, or they may simply be extensions of the rafter.

Roof sheathing used over the exposed tails must be thick enough to accept roofing nails without having them penetrate the exposed side. When ½″-thick plywood is used for roof sheathing and ¾″-thick plywood or other material is used over the exposed tails a wedge ¼″-thick at the butt end is placed on each rafter to feather out the differences in material thickness (see Fig. 7–5).

The open cornice may or may not utilize a fascia board and rain gutter, but spaces between the rafter tails at the building line must be closed off with a trim board to prevent the

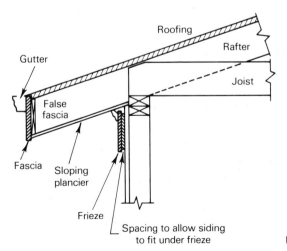

Fig. 7-2   Box Cornice with Sloping Plancier

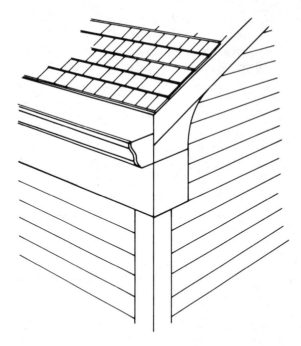

Fig. 7-3   Snub Cornice

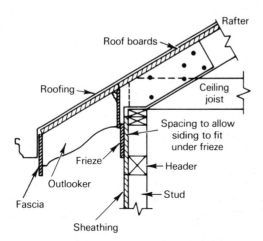

Fig. 7-4   Open Cornice

passage of wind, and so on into the attic area and also to give the building a finished appearance at that point.

Gable end trim usually consists of a rake molding and a rake board, but many variations are possible. The roof may project beyond the end of the building at the gable end and require lookouts, rake board, rake molding, gable plancier, gable frieze, and various moldings (see Fig. 7-6).

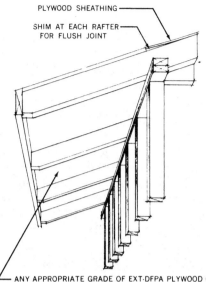

PLYWOOD SHEATHING

SHIM AT EACH RAFTER
FOR FLUSH JOINT

ANY APPROPRIATE GRADE OF EXT-DFPA PLYWOOD OF
ADEQUATE THICKNESS TO PREVENT
PROTRUSION OF ROOFING NAILS OR STAPLES
AT EXPOSED UNDERSIDE, AND TO CARRY DESIGN ROOF LOAD

Fig. 7-5   Compensating for Change in Thickness of
Roof Sheathing

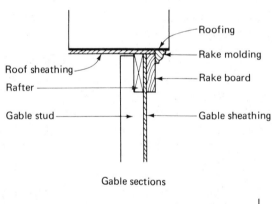

Roofing

Rake molding

Roof sheathing

Rake board

Rafter

Gable stud

Gable sheathing

Gable sections

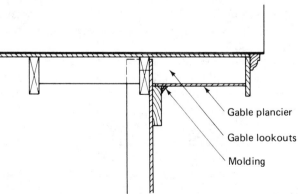

Gable plancier

Gable lookouts

Molding

Fig. 7-6   Gable End Trim

199

Attic Ventilation

All attic spaces need ventilation. To provide this ventilation vent screens are installed at intervals along the length of the plancier and at the top of the roof. Louvers are installed in the gable ends to provide needed ventilation. This ventilation is needed in the summer to allow heat to escape from the attic, and in the winter to allow the flow of air to carry water vapor, which passes through the ceiling material, to the outside and prevent its condensing on the underside of the roof sheathing.

Most building codes have minimum requirements for attic ventilation. The Federal Housing Administration code for one- and two-family structures requires attic ventilation to be not less than $\frac{1}{150}$ of the ceiling area and that at least 50% of the required ventilation be provided in the upper portion of the roof. There are exceptions to this rule, but generally this requirement simply represents good construction practice.

CORNICE MATERIALS

Lumber species used to build the cornice will vary with the function of the cornice member, with local practice, and with what lumber species are locally available.

Parts of the cornice which are not exposed to the weather are usually made of common framing lumber such as southern yellow pine, red fir, white fir, hemlock, or other species locally available. Lookouts, false fascias, and blocking are in this category. The lumber used for these cornice members should generally be straight and free from defects which would weaken them.

Lumber in exposed cornice parts should be dimensionally stable, have good paint-holding characteristics, and be generally weather resistant. Some of the commonly used species which have these characteristics are redwood, cedar, and various western and northern pines.

Redwood is the most weather-resistant and decay-resistant lumber commercially available. It is dimensionally stable and has good nail-holding qualities. Because it does not contain resins and oils, it has excellent paint-holding qualities. It will take and hold almost any finish commonly used on exterior trim. The lack of oils and resins within the lumber also gives it a fire-resistant quality not found in most other lumbers.

There are many species of cedar. Generally they are among the most decay-resistant lumber species available in the United

States. Most cedar is nonresinous, and it is dimensionally stable. This gives it good paint-holding characteristics. Cedar lumber has a straight and uniform grain. It will take nails easily if care is exercised in nail selection. Nails with sharp points spread the grain and when placed near the end of a piece of lumber will cause splitting. Therefore nails with blunt points are recommended.

Most western and northern pines are straight grained, uniformly textured, and dimensionally stable. These characteristics give them good paint-holding ability. They are easy to work with, have good nail-holding power, and good resistance to splitting. While not as resistant to decay as cedar or redwood, on exterior trim they will give service equal to that of cedar or redwood if properly maintained.

Specially grooved fascia boards are available in many areas. These boards may be double grooved or single grooved as shown in Fig. 7-7. The double groove allows the contractor to use the fascia for either ¼"- or ⅜"-thick plancier material by putting the proper groove at the lower edge.

Plywood is used extensively for cornice planciers or soffits. Only exterior-grade plywood should be used as this is the only type which can be expected to resist the effects of weather over a long period of time. Most plywood used for planciers is either ¼" or ⅜" thick, but other thicknesses are occasionally used. Standard 4' by 8' sheets may be cut to the proper width on the job site, or widths of 12", 16", 24", 30", or 36" by 8' in length may be purchased from many local suppliers.

Hardboard of various types, usually ¼" in thickness, is used for planciers, and at least one manufacturer has a complete system of accessories which can be used to simplify and speed the installation of hardboard planciers. The system also provides a neat finished appearance which is durable and easy to maintain.

Moldings of various types are used in cornice work (see Fig. 7-8). Most moldings are stock items manufactured from pine lumber, but it is possible to special order many of them in any lumber species commonly available.

Rake moldings may vary somewhat in shape among different manufacturers, but most of them are 2 ¼" wide by 1 ⅛" thick at the heavy edge. They are used almost exclusively along the rake of the roof on gable ends.

Bed moldings, crown moldings, and cove moldings are available in various sizes. They are used for decorative purposes and to cover the joint between cornice members. The size of

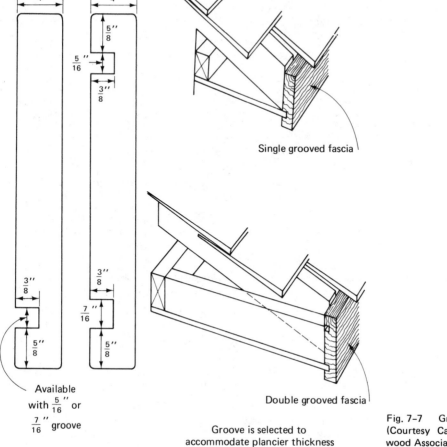

Single grooved fascia

Double grooved fascia

Groove is selected to
accommodate plancier thickness

Available
with $\frac{5}{16}$" or
$\frac{7}{16}$" groove

Fig. 7-7   Grooved Fascia
(Courtesy California Redwood Association)

these moldings is given by the minimum size of the piece of lumber needed to manufacture it. Size can usually be determined by measuring the thickness and width across the corners.

Fasteners used in cornice work are almost always a variety of different sizes and types of nails. Common steel nails and cement-coated box nails are used in areas where they will be completely enclosed and protected from the weather. The nailing of lookouts to the building and rafter tails and the fastening of the false fascia to the rafter tails are places where plain or cement-coated steel nails may be used.

Hot-dipped galvanized nails are commonly used to fasten any cornice member which is exposed to the weather. Nails

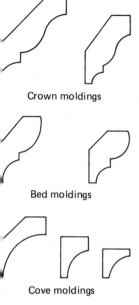

galvanized by the hot-dip process are superior to nails galvanized by methods which leave an acid residue on the nails. The acid residue reacts with moisture in the wood to cause deterioration of the wood. Therefore it is highly recommended that only nails galvanized by the hot-dip process be used to fasten cornice members. These nails should be of sufficient length to hold the material in place and may be of any of the types commonly manufactured (see Fig. 7-9).

Aluminum and stainless steel nails are available for use on exterior trim. Aluminum nails are more costly per pound, but a pound of aluminum nails contains approximately two and one-half times as many nails as a pound of steel nails. Therefore the cost per nail may be less than that for steel nails. Aluminum nails may not be satisfactory for use in some industrial environments.

Stainless steel nails are used on highest-quality work where the added cost of the nails is incidental. Quality stainless steel nails are not affected by any environment.

Crown moldings

Bed moldings

Cove moldings

Fig. 7-8 Moldings Commonly Used in Cornice Work

BUILDING A CORNICE

Various methods are employed to build a cornice. These methods vary with local practice, carpenter preference, and the amount of help available at the job site. The sequence of building the cornice will also vary with the design of the cornice, but there are two general methods of cornice construction. In one method the cornice is built after the walls and roof of the building are erected, and in the other the cornice members are installed before the walls are erected.

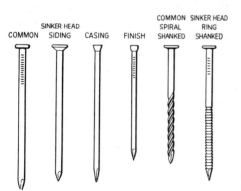

Fig. 7-9 Types of Nails (Courtesy California Redwood Association)

When the cornice is built after the walls are erected it is necessary for the carpenter to build scaffolding from which to work. If the building has a gable end the trim at the gable end is installed first. This involves installing the necessary blocking to accommodate siding or masonry veneer (see Fig. 7-10). The rake board is placed over the blocking and fastened temporarily. The miter cut should be centered at the ridge. After the rake boards are fitted at the ridge with a good tight miter they may be fastened securely to the blocking provided. If the bottom edges of the rake board fail to line up when fitting the board at the miter, it is due to one board running beyond the center of the ridge. By adjusting the boards to the exact center the edges of the rake boards can be brought into alignment.

The rake molding is installed with the top edge flush with the top of the roof boards, with a miter joint where moldings meet at the ridge. The ends of the rake board and molding are usually allowed to run beyond the cornice line at the lower end and are cut to length after the fascia board is installed.

The false fascia is fastened to the rafter tails at the required height in relation to the top of the window frames. Care must be taken to keep the false fascia straight from one end of the building to the other. The false fascia provides backing for the plancier, and the plancier will be only as straight as the false fascia.

Lookouts provide support for the plancier. They are fastened to a lookout ribbon on the building and to the false fascia or rafter rails. Lookouts must be provided at each joint in the plancier material.

The plancier may be installed after the lookouts are in place. To aid in getting a straight line along the fascia edge of the plancier a line is stretched along the false fascia from one end of the building to the other. The plancier is held along this line and nailed to the lookouts, the lookout ribbon, and the false fascia.

The fascia is installed over the false fascia after the plancier is in place. It is nailed to the rafter tails and covers the joint between the plancier and the false fascia. It also provides a finished appearance and a support for the rain gutter or eaves trough.

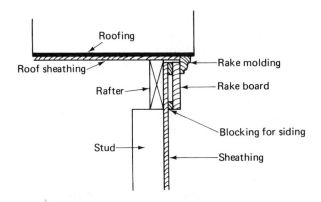

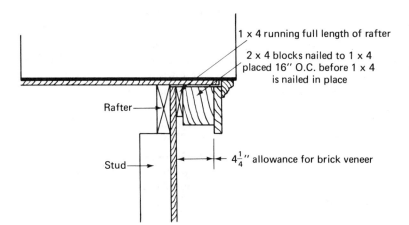

Fig. 7–10   Gable Trim Details

Blocking is necessary to hold the frieze board far enough away from the building to allow for the thickness of siding or masonry veneer to fit behind the frieze. This blocking may take the form of ¼″- or ⅜″-thick filler strips for siding, or it may be 1 by 4 boards with short blocks of 2 by 4 nailed to it every 24″. These short blocks receive the frieze board and should be kept up at least 1″ from the lower edge of the frieze to allow the brick to fit behind the frieze.

The frieze board is fastened over the blocking and nailed securely. A molding which covers the joint between the frieze and plancier is installed, and other moldings required for decorative purposes are put in place.

When the cornice is built before the walls are erected a different sequence is followed in the installation of the various members.

The entire gable end wall is built, and the blocking, rake boards, and rake molding are installed before the wall is erected. Installing the gable trim before the walls are erected is advantageous because it does away with the need for scaffolding. Otherwise the installation of gable end trim is similar to that on walls which are already set in place.

Cornice work on the side walls begins with the location of the lookout ribbon after the sheathing is in place. A chalk line is snapped on the sheathing, and the lookout ribbon is nailed to the studs and along the chalk line.

The location of the lookouts is marked on the ribbon, and lookouts of the proper length are toenailed to the ribbon. The false fascia is nailed to the outer ends of the lookouts, and the lookouts are braced temporarily from the upper side of the cornice.

A string used to align the plancier is stretched from one end of the false fascia to the other. The plancier is aligned with the string and nailed to the lookouts and the false fascia. After the plancier is completed it is held square to the wall and braced from the under side at about 8′ intervals. The finish fascia may be installed at this point, or it may be installed after the walls are erected.

The frieze board and all necessary moldings may be installed after work on the plancier is completed. In most cases it is necessary to finish the ends of the plancier, frieze, fascia, and so on after the walls are erected.

When siding is applied to the building exterior the frieze must be installed in a manner which will allow the siding to fit behind it (see Fig. 7–11). If a masonry veneer is used the frieze is held away from the building 4 ¼″ to 4 ½″ by means of blocking. The space over the window frame must be filled in with some sort of finish material. Generally a 1 by 4 nailed to the top casing of the frame is sufficient (see Fig. 7–11).

DOOR FRAMES

There are many types of door and window frames which the carpenter is called upon to install, but whatever the size or type the same general procedure can be applied for their installation.

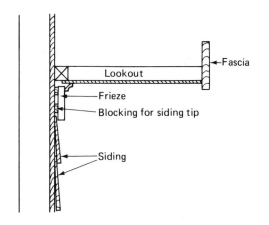

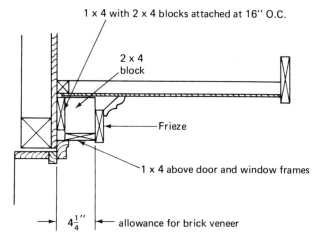

Fig. 7-11   Cornice Details

The parts of a typical door frame are illustrated in Fig. 7-12. The side and head jambs are made of pine and are rabbeted to receive the entrance door. The width of the jambs will vary to accommodate the various thicknesses of the walls. Jambs which are wider than the finished wall thickness can be made usable in many cases by installing a filler strip behind the casing to bring the frame out the proper distance while allowing the frame to project in beyond the rough framework an amount equal to the thickness of the finish wall material.

Door frames with wood sills are installed with the top of the sill at finish floor height (see Fig. 7-13). This is usually accomplished by placing a piece of lumber the thickness of the finish floor at the sill as a height gauge. The sill is leveled with a carpenter's spirit level, and the frame is fastened to the building by nailing through the casing with 16d casing nails. Nails are first placed to hold the sill and head jamb level. Then the frame is plumbed and fastened at the other end of the casing to hold it in a plumb position. Before the frame is fastened at intermediate points along the casing it is straightened by holding a straight edge along the casing and aligning it to the straight edge.

Frames for Sliding Glass Doors

The installation of sliding glass door frames requires special care due to their larger size. There are many different types of sliding glass door frames, but they are somewhat similar in de-

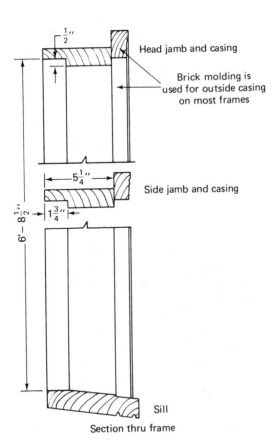

Head jamb and casing

Brick molding is used for outside casing on most frames

Side jamb and casing

Sill

Section thru frame

Fig. 7-12   Typical Door Frame

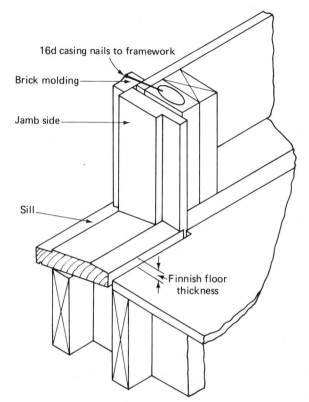

16d casing nails to framework

Brick molding

Jamb side

Sill

Finnish floor
thickness

Fig. 7–13   Door Frame Installation

sign. A typical sliding glass door frame is illustrated in Fig.
7–14.

Care must be taken to install this type of frame with the
sill level and fully supported to prevent sagging under the
weight of the doors. The side jambs and head jambs should be
carefully blocked and securely nailed to prevent movement.

Safety Glass in Sliding Doors   Because of the many injuries
caused by people walking into large glass doors, many com-
munities have passed laws requiring the use of safety glass in
sliding patio doors. Whether required by the local code or not
it is desirable to glaze this type of door with tempered safety
glass or some other safety glazing material to avoid the possibility of serious injury to anyone walking into a closed door.

Fig. 7-14   Sliding Patio Door (Courtesy Andersen Corporation)

FRAME WALL

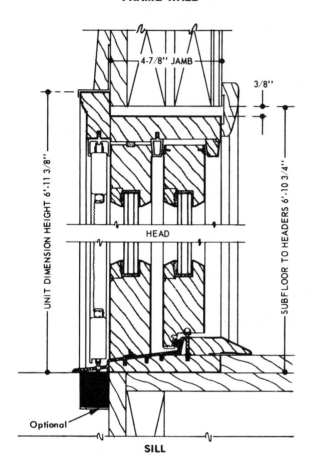

FRAME WALL

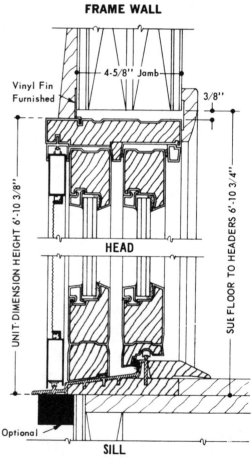

WINDOW FRAMES

There are many types of window frames manufactured by a number of large and small millwork companies. Frames made by different manufacturers have varying convenience features, various types of hardware, and differing construction features. Most window frames come as units with preglazed sashes already installed. Many are available with insulating glass and screen and storm units which are prefitted and ready for installation.

Window frames or window units may be divided into categories according to the manner in which the sash operates, and often different types of units are combined to make a large window unit.

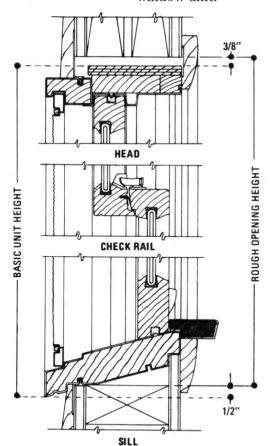

Fig. 7-15   Double Hung Windows (Courtesy Andersen Corporation)

Double-Hung Windows

The double-hung window is one of the oldest and most popular types in use. The sashes in this type of unit are usually balanced with some type of spring balance and bypass each other in a vertical direction (see Fig. 7-15). Two or more double-hung units may be placed together to form a large unit. The space between the two individual units is called a mullion and may vary in width from 2″ to about 6″.

Glide-By Windows

The glide-by window has features similar to the double-hung window, but the sashes bypass each other in a horizontal direction (see Fig. 7-16). Because glide-by windows operate on horizontal tracks it is not necessary to provide the unit with sash balances. This type of unit can be installed at any location in a house, and the smaller sizes are often used in bedrooms and bathrooms because they afford a degree of privacy not available with other types of windows having the same glass area.

Casement Windows

The sash of a casement is hinged on one of the stiles in a manner which allows the sash to swing out. Screens for this type of casement unit are placed on the room side rather than outside as with the double-hung and glide-by windows (see Fig. 7-17). Casement windows should be installed only when the outswinging sash will not be a safety hazard on the outside of the building. Locations where the sash swings out over a walk or driveway are hazardous unless the bottom of the sash is at least 7′ above the grade.

Some older casement units were made to allow the sash to swing into the building. This type of unit has lost favor because it was difficult to seal against rain, and the inswinging sash was always in the way of curtains and furniture when opened.

Awning Windows

Awning window units have sashes hinged on the top rail with operating hardware on the bottom rail which allows the

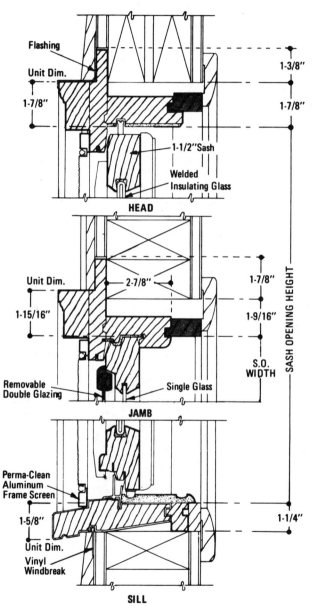

Flashing

Unit Dim.

1-7/8"

1-3/8"

1-7/8"

1-1/2" Sash

Welded
Insulating Glass

HEAD

Unit Dim.

1-15/16"

2-7/8"

1-7/8"

1-9/16"

S.O.
WIDTH

SASH OPENING HEIGHT

Removable
Double Glazing

Single Glass

JAMB

Perma-Clean
Aluminum
Frame Screen

1-5/8"

1-1/4"

Unit Dim.

Vinyl
Windbreak

SILL

Above detail shows stock unit installed in frame wall with ½" sheathing, lath and plaster inside finish.

Fig. 7-16 Glide by Window (Courtesy Andersen Corporation)

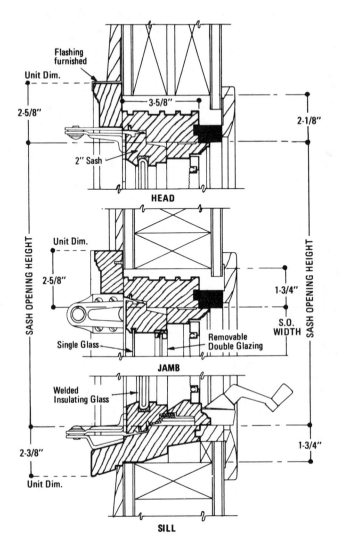

Fig. 7-17   Casement Window (Courtesy Andersen Corporation)

window to swing out at the bottom. This type of window unit is often used to provide ventilation when combined with a larger fixed window unit (see Fig. 7–18).

Hopper Windows

Hopper window units have the sash hinged on the lower rail. Operating hardware allows the sash to swing inward. As

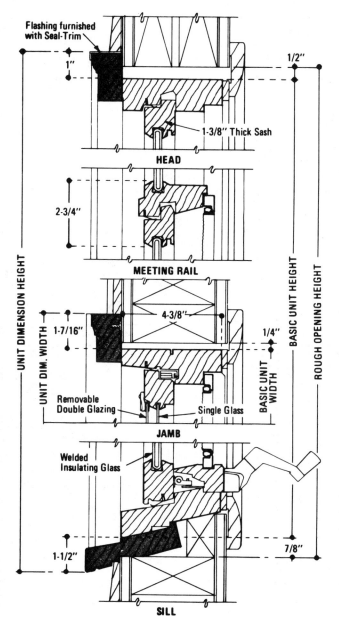

Fig. 7-18 Awning Window Combined with Fixed Sash (Courtesy Andersen Corporation)

215

with awning windows, hopper window units are often combined with fixed window units and are used to provide ventilation (see Fig. 7–19).

Fixed-Sash Windows

Small fixed-sash units are used in window frames installed for decorative purposes. These windows may be round, half-

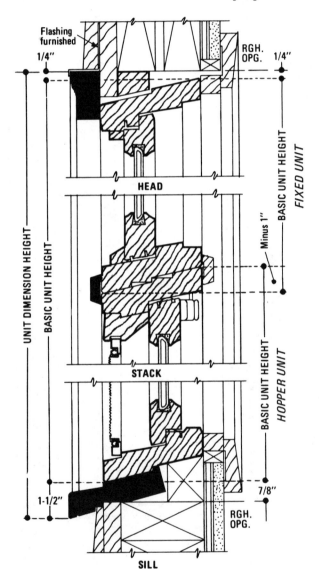

Fig. 7–19 Hopper Window Combined with Fixed Sash (Courtesy Andersen Corporation)

circle, hexagonal, or other shape. Fixed-sash windows are also used in large picture window units (see Fig. 7–20). They may be glazed with ¼″ plate glass, or they may be glazed with insulating glass. Some smaller fixed-sash units are glazed with heavy sheet glass or double-strength window glass.

Fixed-sash frames are often combined with double-hung, casement, awning, or hopper window units. When this is the case the fixed sash provides the picture window area while the operating units provide for ventilation.

Some large window frames are made to accept 1″-thick insulating window units without a sash (see Fig. 7–21). Frames of this type must be sturdily built and rigidly installed to provide adequate support for the insulating unit, which will weigh approximately 7 lb/sq. ft.

Advantage of Wood Windows

Windows for residential construction may be made from many materials. Some of the commonly used materials are enameled steel, aluminum, and wood of various species. Wood window units are the most commonly used and are the best type for a number of reasons.

Wood windows cut air infiltration through efficient weather stripping built in at the factory or installed at the job site. Wood windows, precision machined and treated with a wood preservative, operate easily in all weather and are not

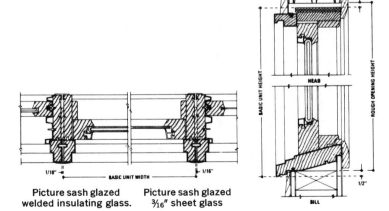

Picture sash glazed welded insulating glass.

Picture sash glazed ³⁄₁₆″ sheet glass

Fig. 7-20   Fixed Sash (Courtesy Andersen Corporation)

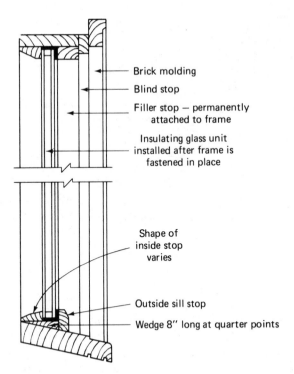

Brick molding

Blind stop

Filler stop — permanently
    attached to frame

Insulating glass unit
installed after frame is
    fastened in place

Shape of
inside stop
    varies

Outside sill stop

Wedge 8" long at quarter points

Fig. 7-21   Frame for Insulating Glass

affected by temperature changes which cause windows made of other materials to expand or contract.

Because wood is a natural insulator, wood window units are more effective at reducing heat loss than metal windows. Wood is approximately 1700 times better as an insulator than aluminum and about 400 times better than steel. Wood window units provide a more efficient control of heat loss when combined with single glazing or insulating glass than either aluminum or steel units.

Wood window units are also effective in helping to control condensation on windows which eventually runs down to damage sills, walls, and furnishings. Because of wood's natural insulating qualities it stays "warm" and prevents condensation on the sash. Condensation may occur on glass during cold weather but can be effectively prevented by using double glazing or insulating glass with wood windows. If metal sashes and frames are used condensation and frost will usually start to accumulate on metal as the temperature drops below 20° F.

Before window frames can be installed they must be prepared for the opening by cutting off the long jamb "ears." These jamb projections should be cut off nearly flush with the top of the head jamb, and they should be cut off on a line squared out from the sill to the rear of the frame (see Fig. 7–22). Cutting the lower jamb ear as illustrated is important, because cutting the jamb flush with the bottom of the sill removes support for the sill. Weight placed on a sill without support would tip the sill and open the joint between the sill and side jamb to the weather.

Window frames are usually installed with the head jamb at door height. That is, the distance from the finish floor to the underside of the head jamb of the window frame will be the same as the distance from the finish floor to the underside of the head jamb in the door frame. Holding the window frame at the proper height before it is fastened in place is often accomplished by use of a height stick (see Fig. 7–23).

The block on the height stick which supports the frame is fastened at a height which holds the frame at the proper distance from the finish floor. The height stick may be made from a 2 by 2 or a 2 by 4. In most cases a 2 by 2 is of sufficient size and is easier to handle.

In some cases the window frame is located by other methods. One of the most common situations in which the height stick is not used is when the window frame is placed tight against the plancier (see Fig. 7–24).

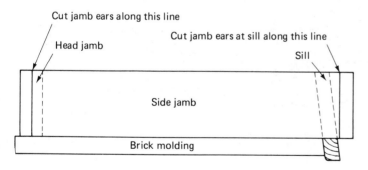

Cut jamb ears along this line

Head jamb

Cut jamb ears at sill along this line

Sill

Side jamb

Brick molding

Trimming jamb ears as shown at sill leaves dado intact to provide a weather tight joint at sill

Fig. 7–22   Preparing Frame for Installation

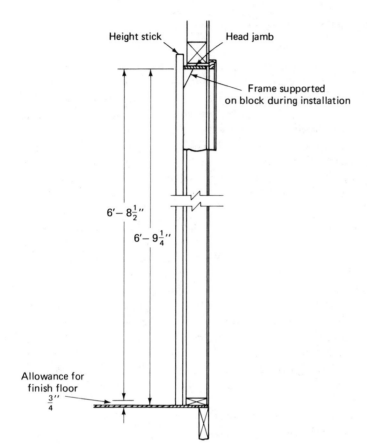

Height stick   Head jamb

Frame supported
on block during installation

$6' - 8\frac{1}{2}''$

$6' - 9\frac{1}{4}''$

Allowance for
finish floor
$\frac{3''}{4}$

Fig. 7-23   Establishing Height of
Window Frame

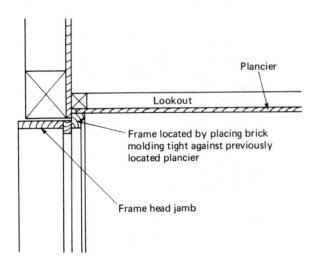

Plancier

Lookout

Frame located by placing brick
molding tight against previously
located plancier

Frame head jamb

Fig. 7-24   Establishing Height of Window
Frame – Alternate Method

Most window frames are fastened to the building by nailing through the outside casing into the studs and rough header with 16d casing nails. Before nailing the frame at one corner it is centered from side to side in the opening. The head jamb (or sill) is then leveled, and the frame is nailed on the opposite corner. Leveling is done by using a carpenter's spirit level at the head jamb or sill, and on large frames a straight edge and spirit level held on the sill are used. The side jambs may be plumbed with a spirit level, or the frame may be squared by measuring the diagonals and adjusting the frame until the diagonals are equal. To completely fasten the frame to the building 16d casing nails are driven in every 12″ to 18″ along the perimeter of the casing. These nails are set below the surface of the casing, and later the holes are filled with putty.

When window frames are installed before the walls are erected they are located along a chalk line snapped across the length of the wall and fastened through the head casing. The frame is squared by measuring the diagonals and adjusting it until the diagonals are equal. It is then completely fastened to the building framework.

Window frames should also be shimmed at the sill, side jambs, and the head jamb, and the shims should be nailed in place. This shimming helps to keep the frames perfectly aligned, thereby ensuring the proper fit and operation of the sash.

EXTERIOR SIDING

Siding on the building exterior gives the building its finished appearance and also serves to weatherproof the structural framework. Therefore the material used for siding must have good weathering characteristics, and it must be able to hold finishes such as paint, stain, or sealers.

Drop Siding

Drop siding is manufactured from many different species of lumber. Among the species commonly used for drop siding are ponderosa pine, Douglas fir, white fir, larch, redwood, and red cedar. The lumber used to make drop siding is often that which is available locally.

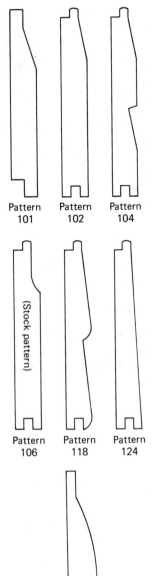

Pattern 101   Pattern 102   Pattern 104

(Stock pattern)

Pattern 106   Pattern 118   Pattern 124

10″ — log cabin siding

Fig. 7-25   Drop Siding

Drop siding is manufactured in various patterns. Some of the available patterns are illustrated in Fig. 7–25. They are identified by the Western Wood Products Association standard pattern number.

Pattern 101 has a shiplap joint and is usually installed over sheathing. The shiplap joint provides an effective seal against the weather.

Pattern 102 has tongue-and-groove edges and is suitable for application directly to the studs. A layer of building paper should be applied to the wall before the siding is installed to help prevent infiltration of air and moisture.

Pattern 104 has a tongue-and-groove edge and can be installed in the same manner as pattern 102. A single piece of pattern 104 gives the appearance of two rows of siding.

Pattern 106 is probably one of the most commonly used drop sidings. The tongue-and-groove edge makes it suitable for application directly to the studs, thereby eliminating the need for sheathing. It has been used extensively on garages, sheds, utility buildings, and residences.

Pattern 118 gives the appearance of two narrow rows of siding with rounded edges. The tongue-and-groove edges make it suitable for installation directly to the studs.

Pattern 124 drop siding gives the appearance of bevel siding when it is installed. Its tongue-and-groove edges make it suitable for installation directly over the studding.

Log cabin siding is available in 6″, 8″, and 10″ widths. It has a shiplap joint, and because of its heavy thickness, 1 ½″ it may be applied directly to the studs.

Application of Drop Siding   The first row of drop siding should be installed along a chalk line which has been snapped on the building. This starting line should be level. Any end joints between sidings should occur over a stud so that they can be fastened securely.

Nails used to fasten drop siding should be hot-dipped galvanized nails of sufficient length to hold securely in the studding. Generally nail penetration of 1″ to 1 ½″ is sufficient. If the siding is used to brace or stiffen the wall greater nail penetration is desirable.

Drop siding is usually nailed with two nails at each stud crossing. However, it is generally desirable to install one nail near the lower edge at each stud crossing. This allows each piece of siding to move slightly as it swells and shrinks because of changes in moisture content (see Fig. 7–26).

Using a single nail through the heavy edge of the siding allows for movement without splitting

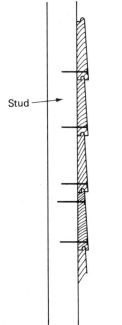

Stud

Generally drop siding is fastened with two nails at each stud crossing. This nailing method braces the wall against wind load

Fig. 7-26   Nailing Drop Siding

Corner treatment for drop siding will vary with the pattern of the siding and the desired finish appearance. The inside corners on all patterns of siding are made with a corner strip (see Fig. 7-27). This corner strip may be ¾″ by ¾″ or 1 ⅛″ by 1 ⅛″. It is generally installed in the corner before the siding is applied to the wall. The siding is fitted to the corner strip, and the end joints may be caulked after the siding is given a coat of finish.

An alternate method of finishing inside corners is to apply the corner strip over the siding after the siding has been installed. This method of inside corner strip application is satisfactory only when the siding is comparatively flat. It generally results in a job which lacks somewhat in finished appearance.

The outside corner of all types of drop siding may be finished with corner boards (see Fig. 7-27). These corner boards may be made from any width material, but 1 by 4 and 1 by 6 boards are most commonly used when the corner board is applied over the top of the siding. To make the corner boards appear equal in width from both directions one board should be cut narrower than the other by the thickness of the board.

When corner boards are installed before the siding is in place they must be made of material which is thicker than the siding. Lumber 1⅛″ thick may be used, or the nominal 1″ boards may be installed over ⅜″-thick shims (see Fig. 7-27). Siding is fitted carefully to the corner boards. The joint between the corner board and the siding is usually caulked after a coat of finish is applied.

Metal siding corners are available for some types of drop siding. They are easy to install after the siding on the adjacent walls is in place. These metal corners offer the appearance of a mitered corner. When painted they blend in with the siding.

The outside corners on some types of drop siding are best finished by mitering. Log cabin siding is one type that looks best when outside corners are mitered. Because of the siding's heavy thickness a portable power saw can be used to advantage to make the miters. This is done by drawing lines on back of the siding square to the edge and following these lines with the saw set at 45°. Siding with mitered corners should be nailed securely to the corner post of the building to prevent opening of the mitered joint as a result of swelling and shrinking of the lumber. Avoid nailing through the miter into the joining siding, since this practice may lead to splitting of the mitered ends.

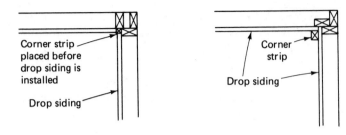

Corner strip placed before drop siding is installed

Drop siding

Corner strip

Drop siding

Inside corners

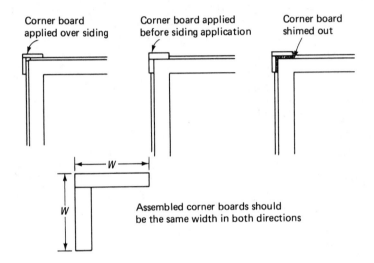

Corner board applied over siding

Corner board applied before siding application

Corner board shimed out

W

W

Assembled corner boards should be the same width in both directions

Outside corners

Fig. 7-27   Corner Finishing

Bevel Siding

Solid wood bevel siding is commonly manufactured from western red cedar and redwood lumber. It is also made from other species of lumber in varying quantities, but these varieties are not as widely distributed as cedar and redwood.

Some of the commonly available sizes of bevel siding are listed in Table 7-1. The cross sections of bevel siding and rabbeted bevel siding are illustrated in Fig. 7-28.

Rabbeted bevel siding is manufactured in many of the same widths and thicknesses as plain bevel siding. The rabbet in the butt edge is usually ½″ wide, and the thickness of the

Using a single nail through the heavy edge of the siding allows for movement without splitting

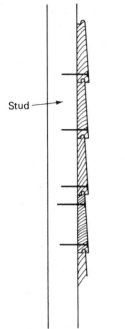

Stud →

Generally drop siding is fastened with two nails at each stud crossing. This nailing method braces the wall against wind load

Fig. 7-26   Nailing Drop Siding

Corner treatment for drop siding will vary with the pattern of the siding and the desired finish appearance. The inside corners on all patterns of siding are made with a corner strip (see Fig. 7-27). This corner strip may be ¾″ by ¾″ or 1 ⅛″ by 1 ⅛″. It is generally installed in the corner before the siding is applied to the wall. The siding is fitted to the corner strip, and the end joints may be caulked after the siding is given a coat of finish.

An alternate method of finishing inside corners is to apply the corner strip over the siding after the siding has been installed. This method of inside corner strip application is satisfactory only when the siding is comparatively flat. It generally results in a job which lacks somewhat in finished appearance.

The outside corner of all types of drop siding may be finished with corner boards (see Fig. 7-27). These corner boards may be made from any width material, but 1 by 4 and 1 by 6 boards are most commonly used when the corner board is applied over the top of the siding. To make the corner boards appear equal in width from both directions one board should be cut narrower than the other by the thickness of the board.

When corner boards are installed before the siding is in place they must be made of material which is thicker than the siding. Lumber 1⅛″ thick may be used, or the nominal 1″ boards may be installed over ⅜″-thick shims (see Fig. 7-27). Siding is fitted carefully to the corner boards. The joint between the corner board and the siding is usually caulked after a coat of finish is applied.

Metal siding corners are available for some types of drop siding. They are easy to install after the siding on the adjacent walls is in place. These metal corners offer the appearance of a mitered corner. When painted they blend in with the siding.

The outside corners on some types of drop siding are best finished by mitering. Log cabin siding is one type that looks best when outside corners are mitered. Because of the siding's heavy thickness a portable power saw can be used to advantage to make the miters. This is done by drawing lines on back of the siding square to the edge and following these lines with the saw set at 45°. Siding with mitered corners should be nailed securely to the corner post of the building to prevent opening of the mitered joint as a result of swelling and shrinking of the lumber. Avoid nailing through the miter into the joining siding, since this practice may lead to splitting of the mitered ends.

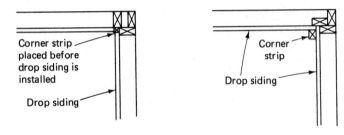

Inside corners

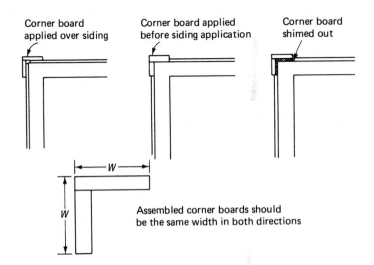

Outside corners

Fig. 7–27   Corner Finishing

Bevel Siding

Solid wood bevel siding is commonly manufactured from western red cedar and redwood lumber. It is also made from other species of lumber in varying quantities, but these varieties are not as widely distributed as cedar and redwood.

Some of the commonly available sizes of bevel siding are listed in Table 7–1. The cross sections of bevel siding and rabbeted bevel siding are illustrated in Fig. 7–28.

Rabbeted bevel siding is manufactured in many of the same widths and thicknesses as plain bevel siding. The rabbet in the butt edge is usually ½″ wide, and the thickness of the

Table 7-1. Bevel siding—common sizes

NOMINAL SIZES (inches)	THICKNESS		WIDTH	MAXIMUM EXPOSURE TO WEATHER
	TIP	BUTT		
$\frac{1}{2}$ by 4	$\frac{3}{16}$	$\frac{15}{32}$	$3\frac{1}{2}$	$2\frac{1}{2}$
$\frac{1}{2}$ by 5	$\frac{3}{16}$	$\frac{15}{32}$	$4\frac{1}{2}$	$3\frac{1}{2}$
$\frac{1}{2}$ by 6	$\frac{3}{16}$	$\frac{15}{32}$	$5\frac{1}{2}$	$4\frac{1}{2}$
$\frac{1}{2}$ by 8	$\frac{3}{16}$	$\frac{15}{32}$	$7\frac{1}{2}$	$6\frac{1}{2}$
$\frac{5}{8}$ by 10	$\frac{3}{16}$	$\frac{9}{16}$	$9\frac{1}{2}$	$8\frac{1}{2}$
$\frac{3}{4}$ by 6	$\frac{3}{16}$	$\frac{3}{4}$	$5\frac{1}{2}$	$4\frac{1}{2}$
$\frac{3}{4}$ by 8	$\frac{3}{16}$	$\frac{3}{4}$	$7\frac{1}{2}$	$6\frac{1}{2}$
$\frac{3}{4}$ by 10	$\frac{3}{16}$	$\frac{3}{4}$	$9\frac{1}{2}$	$8\frac{1}{2}$
$\frac{3}{4}$ by 12	$\frac{3}{16}$	$\frac{3}{4}$	$11\frac{1}{2}$	$10\frac{1}{2}$

Rabbeted bevel siding

Bevel siding   Fig. 7-28   Siding Detail

rabbet is equal to the thickness of the top of the siding. The heavier rabbeted bevel sidings are generally more acceptable because they are more resistant to splitting at the rabbeted edge than the thinner sidings. For this reason some manufacturers produce rabbeted bevel siding only in the ¾″ thickness.

Application of Bevel and Rabbeted Bevel Siding   Beveled sidings are usually installed horizontally. A chalk line snapped on the building locating the top edge of the first course aids in installing the first row of siding. The first row of bevel siding should be installed over a filler strip which takes the place of the siding tip in the succeeding rows (see Fig. 7-29).

Succeeding courses of bevel siding are located in accordance with a story pole. Story poles are usually strips of wood ¾″ by ¾″ or ¾″ by 1 ½″ on which the butt edge of each siding course is marked. The story pole is laid out by first dividing the height of the wall to be covered with siding by the maximum siding exposure to get the number of siding courses. The result nearly always is a number of courses and a fraction. Since all siding courses should be of equal width, the result is rounded off to the next full number. The number of courses is divided into the height of the wall to get the siding exposure, and the exposure is located on the story pole with a divider and marked in with a pencil and combination square.

Most building codes require a minimum lap of 1″ for bevel siding. This is reflected in Table 7-1, which gives the maximum exposure. A greater lap is always acceptable, since bevel siding depends on the lap of succeeding boards for its weatherproofing ability.

Nails used to fasten bevel siding should be made of stainless steel, aluminum, or steel that has been galvanized by the hot-dip process. Generally 8d nails are used for ¾″-thick siding, and 6d nails are used for ½″-thick siding. All siding should be nailed with one nail at each stud crossing. The nail should be driven through the butt edge of the siding in a manner which allows it to pass above the top of the siding course below (see Fig. 7-29). Nailing in this manner allows for free movement in each piece of siding caused by swelling or shrinking in width. This free movement helps to avoid splitting and bulging of the siding. When nailing near the edge or end of the siding it is advisable to predrill the nail or blunt the top of the nail to prevent splitting the wood. This is also important when fastening near mitered corners. Nails should be driven up to the face of the siding without leaving any nicks or hammer marks. If the nails are set below the surface of the siding the nail holes should be filled with putty after the prime coat of finish is applied.

Inside corners are made with the siding ends fitting against an inside corner strip which is usually about 1 ⅛″ square.

Outside corners may be finished with corner boards, metal corner caps, or by mitering. When corner boards are used they are installed before the siding is applied, and the siding ends are fitted to the corner boards in a manner similar to that for drop siding illustrated in Fig. 7-27.

Metal corner caps may be purchased for the various widths and thicknesses of sidings. These caps are usually made of galvanized steel, or they may be made of preprimed aluminum. Corner caps are installed after the siding on the adjacent walls is in place. They are usually fastened by nailing through the corner cap on each wall at the same level as the siding is nailed to the building.

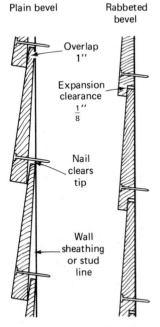

Plain bevel   Rabbeted bevel

Overlap 1″

Expansion clearance $\frac{1}{8}$″

Nail clears tip

Wall sheathing or stud line

Fig. 7-29   Nailing Detail (Courtesy California Redwood Association)

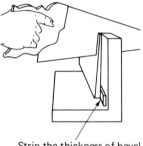

Strip the thickness of bevel siding at exposure width

Fig. 7-30   Mitering Bevel Siding

Mitered corners give a pleasing appearance. Pieces which make up the mitered corners are usually installed first, and the space between the ends of the building is filled in with pieces cut to length. To make the mitered cuts on bevel siding a strip of wood equal in width and thickness to the lap of the siding tip is placed in the miter box (see Fig. 7-30). This strip of wood puts the siding in the same relative position in the miter box that it has when it is placed on the wall. The saw passes through the siding at an angle of 45° in a horizontal plane.

The lap in rabbeted bevel siding is governed by the depth of the rabbet. When this type of siding is installed a clearance of $\frac{1}{16}$" to $\frac{1}{8}$" should be left for expansion between adjacent rows of siding (see Fig. 7-29). Rabbeted bevel siding should be fastened with one nail at each stud crossing. Inside and outside corners are finished in the same manner as for regular bevel siding.

Vertical Board Siding

Shiplap and tongue-and-groove boards with various edge treatments may be used for vertical sidings. These boards are applied vertically because vertical joints provide for natural drainage of rain water.

Various patterns of both shiplap and tongue-and-groove sidings are available. The nominal sizes range from 1 by 4 to 1 by 12. Some of the available patterns are illustrated in Fig. 7-31.

All vertical sidings require some type of backing. This backing should not be more than 24" on center. The best backing is provided by 2 by 4's installed at 2' centers between the studs.

Nailing recommendations for shiplap and tongue-and-groove sidings are given in Fig. 7-32. Nails used should be weather resistant as with other types of siding. When fastening shiplap two nails placed about 1" from the edges are used at each nailing support. Even though regular shiplap is illustrated in the nailing detail this type should be avoided if there is a desire to hide the joints between boards. Joints may be effectively concealed by using material with V-groove or channel-groove edges.

Tongue-and-groove material is fastened by blind nailing.

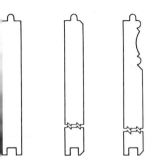

T and *G* siding patterns

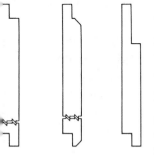

Shiplap siding patterns

Fig. 7-31   Tongue & Groove and Shiplap Sidings

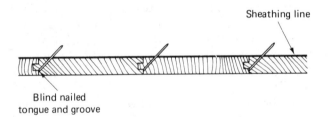

Sheathing line

Blind nailed
tongue and groove

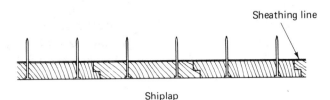

Sheathing line

Shiplap

Note: These sidings may be
installed either horizontally
or vertically

Fig. 7-32   Tongue & Groove and Shiplap Nailing Detail (Courtesy California
Redwood Association)

That is, all the nails except those in the first and last boards are concealed by the groove of the boards. Edge nailing is another term used for blind nailing. The boards are nailed only at the tongue, as illustrated in the nailing detail. One nail is used at each support crossing. As with shiplap, square-edged boards are avoided when there is a desire to hide the joints between the boards. V-groove tongue-and-groove boards are usually used.

Application of Vertical Board Siding   The first board installed should be placed plumb and securely nailed in place. The lower end should lap over the foundation wall by a minimum of 1″, and all succeeding boards should be kept in line at the bottom end. A line may be stretched from one end of the building to the other to aid in aligning the lower end, or a guide board may be fastened in place temporarily and the lower ends of the succeeding boards allowed to set on the guide board. This board is removed after all the siding is in place.

As the succeeding boards are placed they should be checked for plumbness, and adjustments should be made as needed.

Board-and-Batten Siding

There are many possible variations with board-and-batten siding. Some of the more commonly used board-and-batten systems are illustrated in the nailing detail of Fig. 7–33. They are applied with the boards running vertically because the vertical joints afford natural drainage of rain water.

Standard board and batten

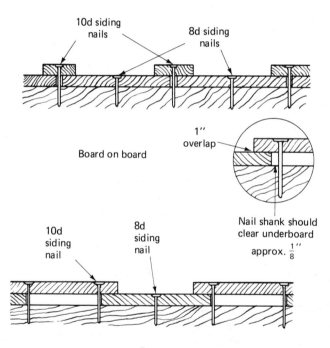

10d siding nails

8d siding nails

Board on board

1″ overlap

Nail shank should clear underboard approx. $\frac{1}{8}$″

10d siding nail

8d siding nail

Reverse batten

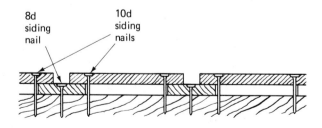

8d siding nail

10d siding nails

Fig. 7–33   Batten Systems – Nailing Detail

The standard board-and-batten system consists of comparatively wide boards fastened directly against the building sheathing with narrower boards or battens covering the joints between the boards. The underboards should be nailed with one nail at each support. Nails should be placed midway between the edges of the board, and all nails in a board should be in line from top to bottom. The battens are nailed over the joints with the nails passing through the space between the batten boards. Nailing in this manner allows the boards to move slightly with changes in moisture content without causing splitting or cupping.

The board-on-board system used wide boards in place of batten strips. The underboard is fastened with one nail at each support. These nails are placed midway between the edges of the board. The overboards should lap 1″ on the underboards. Nails in the overboards are placed so that they pass the underboards with about ⅛″ clearance. This clearance allows the nails to move or bend slightly in the event the overboard swells or shrinks as a result of moisture content changes.

The reverse board-and-batten system uses narrow boards for the underboard. The wide overboards give the reverse board-and-batten appearance. Reverse board-and-batten systems are fastened to the building in the same manner as the board-on-board system.

Application of Board-and-Batten Siding   Before any batten boards are installed the carpenter must check the length of the wall and lay out the spacing of the boards and battens. By increasing or decreasing the spaces between the various boards he can make all the battens on a wall appear to be the same width.

The underboards are cut to length and installed plumb in accordance with the carpenter's layout. After the underboards are in place the overboards or battens may be installed. Care must be taken to install these members plumb as they give the wall its finished appearance.

Plywood Siding

All plywood for exterior use is manufactured with waterproof glue, and there are many different patterns available which give a variety of exterior appearances. Plywood for exterior finish use is most commonly manufactured from Douglas

Fig. 7-34 Plywood Textures (Courtesy American Plywood Association)

Texture One-Eleven

Deep grooves cut into face for sharp shadow lines. Grooves 1/4″ deep, 3/8″ wide, 4″ or 2″ o.c., panel thickness 5/8″. Other groove spacings available are 6″ and 8″ o.c.. Standard type has unsanded face with natural wood characteristics for rustic effect. Available with Medium Density Overlaid, sanded, rough sawn, brushed, rough sanded, or striated faces. Finish unsanded and textured surfaces with exterior pigmented stains. Overlaid and sanded surfaces are intended for exterior paint. Available in Redwood, Cedar, Douglas fir, and other species.

Channel Groove

Grooves 1/16″ deep, 3/8″ wide, cut into faces of 3/8″ thick panels, 4″ and 2″ o.c. Other groove spacings available. Ship lapped for continuous patterns. Available in similar surface patterns and textures as Texture One-Eleven. Finishing recommendations are the same as for Texture One-Eleven. Available in Redwood, Cedar, Douglas fir, and other species.

Reverse Board & Batten

Deep, wide grooves cut into brushed, rough sawn, coarse sanded, or natural textured surfaces. Grooves 1/4″ deep, 1-1/2″ wide, spaced 12″ or 16″ o.c. with panel thickness of 5/8″. Provides deep, sharp, shadow line. Long edges are ship lapped for continuous patterns.
Finish with exterior pigmented stain (or leave natural without finish for weathered rustic effect). Available in Redwood, Douglas fir and Cedar.

Rough Sawn

Saw textured surfaces designed to combine the charm of natural wood with ease of installation provided by large panels. Also available in various types of grooved surfaces and in lapped siding widths of 8″, 10″ and 12″, with lengths to 16′. Available primed and unprimed.
The rough unsealed surface is especially suitable for exterior pigmented stain. Available in Redwood, Douglas fir, and Cedar.

Kerfed

Light touch sanded or rough sawn surface with narrow square cut grooves to provide a distinctive effect. Long edges are ship lapped for continuous patterns.
Grooves 1/8″ wide, 1/16″ deep, 4″ o.c. are cut into unrepaired natural solid faces of Douglas fir. Available also in groove spacings of 2″, 6″ and 8″ o.c.
Surface especially suitable for exterior pigmented stain. Available in Douglas fir.

Striated and Corrugated

The fine striations are random width, closely spaced grooves forming a vertical pattern. The striations conceal nail heads, checking and grain raise, and eliminate unsightly joints. Finish with exterior paint or pigmented stain.

fir, redwood, and cedar, but some patterns are made from other species. Figure 7-34 illustrates some of the plywood surface textures available along with a short description of each.

Many of these plywood sheets can be used to give the reverse board-and-batten effect or the appearance of V-groove boards. One of the advantages in using plywood is that the installation of large sheets cuts down on the number of construction joints which can leak water and air. Plywood is dimensionally stable, so there is little need to provide for expansion and contraction.

Recommended joints for large-size panels are shown in Fig. 7-35. Notice that a clearance of $\frac{1}{16}''$ is specified between the edges of adjacent sheets in all cases. Battens over vertical joints of square-edged sheets are not required but are recommended to make the joint waterproof. Vertical shiplap joints effectively seal against the weather, and horizontal shiplap joints are satisfactory in sealing against the weather. Horizontal joints which are flashed provide a better weather seal but offer a slightly less attractive appearance.

Another advantage to using large plywood sheets is that in many cases they can provide bracing, sheathing, and finish siding, all in one application.

Fig. 7-35   Joint Details for Plywood Sidings (Courtesy American Plywood Association)

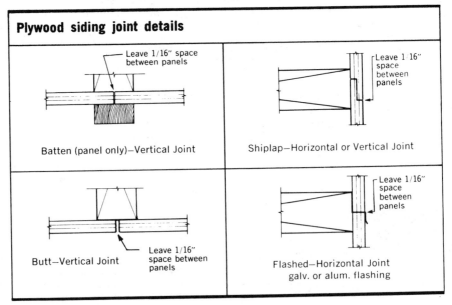

Various types of hardboard sidings are available for residential use. The horizontal lap siding provides the appearance of traditional bevel siding. Large-size sheets which may be installed to give the appearance of boards and battens, V-groove paneling, or grooved paneling are 4' wide and available in lengths of 8' to 12'.

Application of Hardboard Lap Sidings   Hardboard lap sidings should be cut with a fine-tooth saw. End joints between sidings should be closely fitted but not forced together.

A starter strip is used for hardboard lap siding just as for bevel siding and other types of lap siding. This strip is usually ⅜" thick by 1 ½" in width.

Nails used to fasten hardboard exterior sidings should be weather resistant and of the type recommended by the hardboard siding manufacturer. Nails 2 ½" long (8d) are commonly used. They should be placed at least ½" from the edges and ends of the siding and should penetrate both siding courses. Where joints occur nails should be driven through both siding courses (see Fig. 7–36).

The procedure for installing hardboard lap siding is similar to that for installing other types of lap siding. First a starter strip is installed, and then the first course of siding is applied over the starter strip and leveled. Nails are driven along the bottom edge at each stud location. Succeeding courses of siding are aligned according to a predetermined layout and fastened securely in place. Where siding fits against door and window casings there should be a slight space. Hardboard sidings should never be forced or snapped into place. Joints at door and window frames should be caulked.

Application of Hardboard Panel Sidings   Hardboard panel siding may be placed over sheathed or unsheathed stud walls with a maximum stud spacing of 16" O.C. or 24" O.C., depending on the type of hardboard being used. As with other types of hardboard a fine-tooth saw should be used for cutting panel siding.

The joints and edges of all panels should fall on the center of framing members. Butt joints should be covered with batten strips to make them waterproof, and additional batten strips may be installed over intermediate studs for desired appearance. Battens should be installed only where an adequate nailing base is provided.

Fig. 7-36   Hardboard Lap Siding – Nailing Details (Courtesy Masonite Corporation)

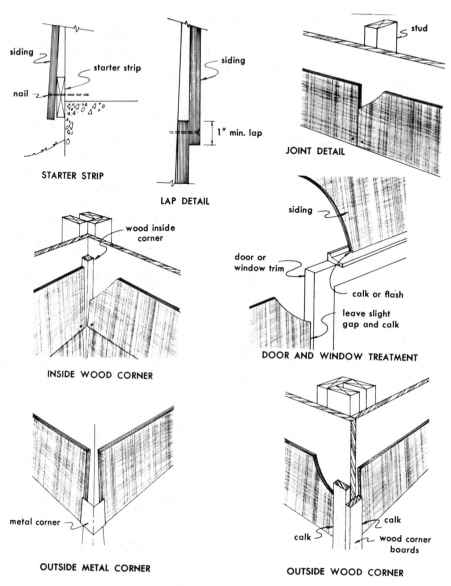

STARTER STRIP

LAP DETAIL

JOINT DETAIL

INSIDE WOOD CORNER

DOOR AND WINDOW TREATMENT

OUTSIDE METAL CORNER

OUTSIDE WOOD CORNER

Hardboard with shiplap edges is applied with the edges at the center of a framing member. If horizontal joints are necessary they should be beveled at 45° and caulked. Various nailing details for hardboard panel sidings are illustrated in Fig. 7-37.

Generally 6d or 8d galvanized nails with a ³⁄₁₆″ head are used to fasten hardboard panel sidings. These nails are usually

Fig. 7-37   Hardboard Panel Siding – Nailing Details (Courtesy Masonite Corporation)

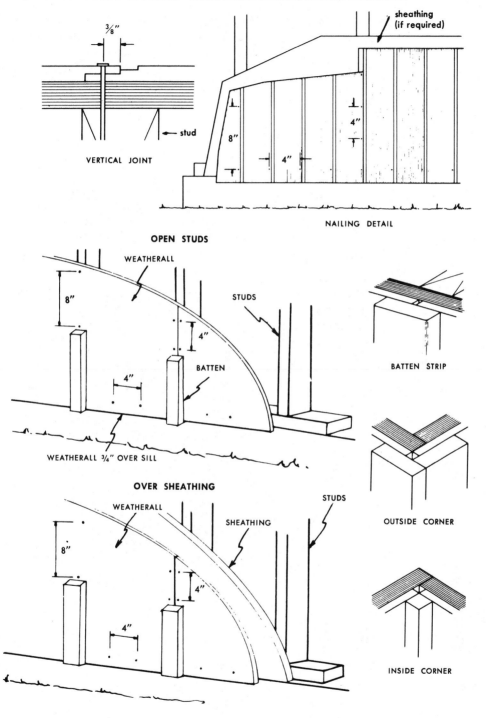

VERTICAL JOINT

NAILING DETAIL

OPEN STUDS

WEATHERALL ¾" OVER SILL

OVER SHEATHING

BATTEN STRIP

OUTSIDE CORNER

INSIDE CORNER

placed ³⁄₈″ to ¹⁄₂″ from the panel edges and are spaced every 4″ to 6″ along the edges. Nails on intermediate supports are spaced 8″ to 12″ apart. The size of nail and nail spacing varies with the type of siding and the type of installation. Manufacturers' recommendations should always be followed when installing hardboard sidings.

Aluminum Siding

Aluminum siding is available in many colors and a number of different widths. It is made in patterns which resemble horizontal bevel siding, vertical V-groove, and board-and-batten siding and is available with a fiberboard insulation bonded to the back.

Application of Aluminum Siding   In many ways the installation of horizontal aluminum siding is quite similar to the procedures followed when applying other types of sidings. A level line must be established, locating the top of the first course of siding. Using the level line as a guide a starter strip is installed along the building wall at the level of the lower edge of the siding. This strip is nailed approximately every 8″ and the ends are kept back from the ends of the wall to allow for siding corners.

Inside corner posts, if needed, must be installed before the siding is put in place. Corner posts should be set square into the corner, aligned, and nailed every 12″ (see Fig. 7-38). If outside corner posts are used in place of siding corners they must also

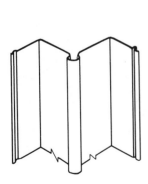

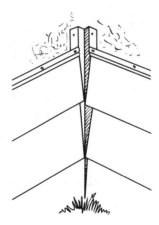

Fig. 7-38   Aluminum Inside Corner Post

be placed before the siding is installed (see Fig. 7–39). When the siding is fitted into the outside or inside corners a $\frac{1}{16}''$ space should be allowed for expansion between the siding and the corner post. The joint is sealed by caulking after the entire wall is completed.

Nails used to apply aluminum siding must be made of aluminum to avoid galvanic action between the aluminum siding and the nails. Always follow the recommendations of the siding manufacturer when selecting nails.

The bottom course of siding is locked to the starter strip and nailed to the building. Extra care must be taken to see that this course is locked securely and aligned properly, since it establishes the base on which all other courses are applied. Nails should be applied every $16''$ along the entire length of the siding. As the siding is nailed in place a check should be made to be sure that it is locked in the starter strip along its entire length (see Fig. 7–40).

Nails should be driven straight in but never angled up or down. They should be placed at the center of the elongated slot to allow the siding to expand and contract with temperature changes. Nails should never be driven in the spaces between slots because this causes dents in the siding edge and makes it difficult to install the succeeding course. Aluminum siding should be "hung" on nails and not nailed tightly to the wall. Nails should be driven to within $\frac{1}{32}''$ of the siding, being careful not to dent the edge of the siding (see Fig. 7–41). Where there are small irregularities in the wall the nails should be driven in only far enough to hold the siding without bending it.

As a general rule aluminum siding is cut with a fine-tooth portable saw set in a jig, or with a radial-arm saw. It can also

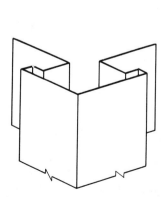
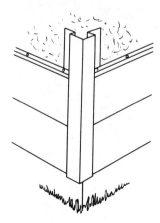

Fig. 7–39   Aluminum Outside Corner Post

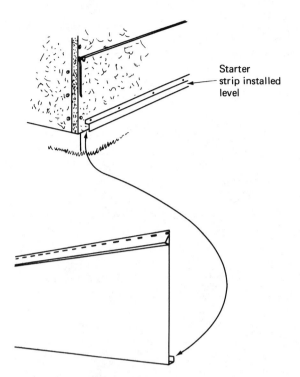

Starter
strip installed
level

Special care should be taken to
be sure lower edge of starting course
is securely locked on the starter strip

Fig. 7–40   Starting First Course

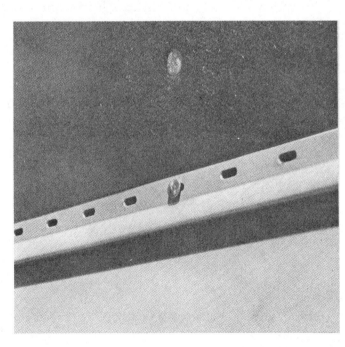

Fig. 7–41   Nailing Detail

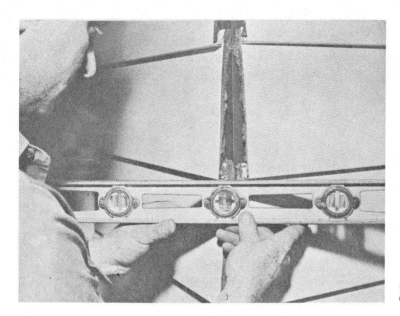

Fig. 7–42   Aligning Siding Courses

be scored with a sharp utility knife and cut with a tin snips after the insulation has been removed from the back.

When planning an aluminum siding job the laps should be made so that they are least visible from the direction of greatest traffic. This is accomplished by lapping the sidings over the top of previously applied sidings when moving toward the direction of greatest traffic. The effect of this procedure is to make the laps nearly invisible from one direction.

When individual corners are used the siding courses on adjacent walls must be at the same level so that the corners will lock into the butt edge (see Fig. 7–42). A 2½″-long nail driven through the hole provided should be used to hold the corner in place. All joints between door and window frames should be caulked with color-matched caulk.

The entire job should be cleaned of dirt and handprints, and all metal scraps and packing should be picked up and disposed of before leaving the job.

REVIEW QUESTIONS

1. What are the purposes of exterior trim?
2. What characteristics must materials used for exterior trim have?

3.   Describe two main cornice types.

4.   List and define the parts of a box cornice.

5.   Why is attic ventilation necessary?

6.   Outline the procedure followed in building a cornice before wall erection.

7.   Make a cross-sectional sketch of a door frame. Name all parts.

8.   Why should safety glazing materials be used in doors?

9.   Name and describe the various types of window frames used in residential construction.

10.   How are window frames fastened to the building framework?

11.   Name and describe the various types of wood siding.

12.   Outline the procedure for installing drop siding.

13.   Outline the procedure for installing bevel siding.

14.   Outline the procedure for installing vertical board siding.

15.   Outline the procedure for installing board-and-batten siding.

16.   Outline the procedure for installing plywood siding.

17.   Outline the procedure for installing hardboard siding.

18.   Outline the procedure for installing aluminum siding.

building services

8

Building services include plumbing, heating, air conditioning, and electrical work. This work is often referred to as the mechanical work. On large jobs elaborate plans are needed to show the location of all the various piping, outlets, fixtures, registers, and radiators. Residential jobs do not require elaborate plans for mechanical work. Simple plans showing the location of fixtures, switches, outlets, and registers are sufficient.

PLUMBING

The plumber is one of the first mechanical tradesmen called upon in residential construction to install some of the necessary building services. He will install the waste stack, various drains, hot water and cold water lines, the bathtub, and any other piping which will be enclosed in the walls in the completed building.

Before the plumber starts work in a building he will study the plans and mark the location of the stack, closet drains, bathtub drain, lavatory drains, kitchen sink drain, and any other drain lines required by the plans and specifications.

After locating all the drains and doing the necessary layout work the plumber will cut the necessary holes through the subfloor, studding, and other framing members as necessary. It is important that the carpenter locate structural members in a manner which will allow the plumber to install the necessary piping without the need for cutting any of them.

The waste stack is installed from the first floor up through the roof. To install this stack it is necessary for the plumber to cut a hole through the floor for each closet bend. He will also cut out a portion of the top and bottom plates of the wall in which the stack is placed. A hole must also be made through the floor for the stack to pass through.

The waste stack is usually made from cast iron pipe, copper pipe, or plastic pipe. Various building codes have requirements to be met for the use of these materials under varying conditions, and it is important that the builder know these requirements and build accordingly to avoid unnecessary delays.

Waste lines and vents for lavatories, bathtubs, and shower stalls are piped into the waste stack. Depending on the location, the kitchen drain may also be piped into the waste stack. If the kitchen sink is located some distance from the waste stack, a separate vent stack and drain are installed for the sink.

Hot and cold water lines for lavatories, sinks, bathtubs, and showers are run in the walls, and all necessary fixtures are installed. Cold water supply lines are also roughed in for the water closets. The water lines are usually run a short distance through the floor. The plumber will connect these pipes to the water supply when he does the finish work in the building.

Plumbing Finish Work

Finish work for the plumber usually includes the connecting of all water supply lines and drains, and the installation of hot water heaters, stationary tubs, and other fixtures.

Also included under finish work is the installation of sinks, food waste disposers, dishwashers, lavatories, water closets, and tub and shower fittings. Kitchen sinks and dishwashers cannot be installed until the necessary cabinets and counter tops are in place. Food waste disposers cannot be installed until the sink is in place. Therefore it soon becomes apparent that cooperation among the trades is necessary to complete the job in a reasonable amount of time.

Bathtub and shower fittings cannot be installed until the ceramic tile, if used, is installed, and water closets cannot be installed until the finish floor which will support them is completed. Plumbing fixture installation, therefore, is one of the last items to be installed in the building.

HEATING AND SHEET METAL

The sheet metal worker is usually the first of the mechanical tradesmen on the job. He installs rain gutters and all necessary flashing and counter flashing around chimney and other places as needed. Rain gutters are usually installed before the roofing is applied. (If wood gutters are used they are installed by the carpenters.) Any counter flashing required cannot be installed until the roofing is completed.

If the building will have a hot air heating system the sheet metal contractor will locate the hot air registers and the cold air returns after he completes the gutter work. He will then proceed to do the necessary rough-in work for these registers.

Cold air registers usually require that a portion of the bottom plate, subfloor, and wall be cut away. A sheet metal duct is installed in the opening according to the size of the register. The lower end of the duct passes through the floor, where it will later be connected to ductwork leading to the furnace.

Hot air baseboard registers in a single-story residence usually require only that a hole be cut through the floor between the joists at the finish wall line. Later hot air pipes from the furnace are run to the opening and connected to the baseboard registers. The registers cannot be installed until the finish wall and finish floor are completed.

Hot air registers installed above the floor require duct work placed between the studs. This duct work is roughed in by cutting the plate and floor away between the studs and then installing the duct so that it projects a short distance below the floor, with the top set at register height.

Because of the space taken by these ducts, it should be apparent to the carpenter that care must be taken in the layout of stud and joist location so as to provide the necessary room for heating ducts.

Heat ducts for a second floor are installed in a manner similar to that for registers above the floor. The double top plate is also cut out and the duct is run up to the space between the floor joists. There it may be turned to run parallel with the floor joists until it comes to the location of the hot air register. At that point the necessary holes are cut and the duct work is installed through the floor as required.

Cold air returns on the second floor require duct installation identical to that on the first floor. The space between the joists and studs which is enclosed by the interior walls functions as a duct. The heating contractor simply makes sure that there are no obstructions between the joists or studs which would prevent the circulation of air.

The installation of the furnace and connecting duct work is usually done sometime after the building is under a permanent roof. Quite often it is not done until the building is nearly completed.

Hydronic Heating

The rough-in of hydronic heating for a single-family one-story residence usually involves very little work. Any necessary piping which must be enclosed in the walls is installed before the interior wall material is applied, but many hydronic systems do not require any rough-in before the interior walls are complete, because supply pipes are passed through the floor directly below the radiators and all necessary connections are made in the basement.

When a hydronic heating system is installed on a second floor all supply lines must be installed in the walls with necessary branches and controls before the floor and walls are enclosed.

Electric Heat

Electric heating systems of various types are available. The rough-in for each type varies, but in all cases wires which will be enclosed in the finished wall must be installed before the wall material is placed.

ELECTRICAL WORK

All electrical work must be installed in accordance with the local building codes, which also require that the installation be done by qualified, licensed electricians. This requirement is made to guarantee public safety against the hazards of electrical shock caused by improper installations.

As with other building services, electrical work may be divided into two or more parts. Generally in residential work it is thought of as either rough work or finish work.

Rough Work

Electrical rough-in usually begins with the electrician marking the location of the various switches, outlets, and fixtures in accordance with the building plans. After marking the locations, electrical outlet boxes of the proper sizes are fastened in place. Next holes are drilled through the studs, plates, and joists for the electric cables (or conduits) as required to provide the circuits called for by the plan.

Cables are pulled through the holes and connected to the boxes to form the necessary circuits. Feeder lines are run to the wall directly above the electrical service box and passed through the bottom plate and floor to the service box located in the basement.

The electrical service rough-in involves the installation of the main switch box, the conduit to the meter box on the outside of the building, the meter box, and the necessary piping and wiring to the point where the power company will connect the power lines.

Finish Work

Electrical finish work is done after the interior wall material has been installed. At that time switches, plug receptacles, and even ceiling fixtures can be installed. While ceiling fixtures can be installed at the same time receptacles are put in, their installation is usually deferred until the building is nearly complete in order to avoid breakage. Wall plates around switches and receptacles are usually installed after the walls have been painted.

The feeder lines are usually connected to the main service after the switches and receptacles have been installed. This is done as an extra precaution to avoid accidental electrical shock if some of the circuits are activated prematurely.

Other finish work done by the electrician involves connecting diswashers, food waste disposers, and electric water heaters installed by the plumber. He will also provide necessary wiring for furnaces, ranges, dryers, and even door chimes.

The installation of electrical work requires the coordination of the work of a number of different trades. If the job is well planned and if each trade is scheduled with proper lead time the entire project can be completed with few delays.

REVIEW QUESTIONS

1.   What are building services?
2.   What work is included in rough plumbing?
3.   When is rough plumbing installed?
4.   What work is included in finish plumbing?
5.   What work comes under the heading of sheet metal?
6.   How are warm air heating systems roughed in?
7.   When is electrical rough work done?
8.   When is electrical finish work done?

insulation

9

Insulation is used to control the passage of sound, heat, and humidity through floors, walls, and ceilings. Therefore it may be said that insulation is used to control comfort. To provide comfort each room requires the right combination of temperature and humidity, and it must be shielded from outside noises.

The temperature required to provide comfort will depend on the amount of activity, the relative humidity of the air in the room, and personal preference. Most people feel comfortable when indoor temperatures are between 68°F degrees and 74°F.

When there is a temperature change of 3° in a period of 1 hour most people will feel some discomfort, and when wall temperatures drop to 3° or more below the inside air temperature a person's body radiates heat to the walls. This loss of heat creates a chilly feeling.

The proper amount of insulation placed in walls, floors, and ceilings can prevent discomfort and provide savings in heating and cooling costs. The type and amount of insulation and the manner in which it is installed determines its effectiveness. Therefore different methods of installation and different materials are recommended to fill the various needs for insulation.

Insulation is available in several forms and may be placed in categories according to the form in which it is supplied.

Rigid Insulation   Rigid insulation is supplied in 2' by 4' sheets for use on flat roofs and in 2' by 8' sheets for use on walls or floors. It may be ½" or ²⁵⁄₃₂" thick. It is similar to building sheathing in appearance and is made from various materials. Some of the materials used to manufacture fiberboard rigid insulation are wood fiber, sugarcane, and various vegetable fibers. Rigid insulation made of polystyrene and urethane plastic foams are available in thicknesses from ½" to 2" but these are not often used in residential construction.

Blanket Insulation   Blanket insulation is furnished in rolls containing from 70 to 153 sq. ft. The blankets may be obtained in widths of 15" or 23" to accommodate the 16" or 24" spacing of framing members. Some insulation products are available in 11" and 19" widths to meet special conditions. The thickness of the blanket is usually between 2" and 4". Blanket insulation may be faced with an asphalted kraft paper or aluminum foil. Flanges on this facing are used to fasten the insulation to the framework and to provide a vapor barrier. This type of insulation is usually manufactured from glass fiber or mineral wool.

Batt-Type Insulation   Batt-type insulation is similar to blanket insulation except that it is furnished in packages containing pieces from 48" to 96" long. Batts are available in thicknesses from 2" to 6". They may be obtained with kraft paper facing, aluminum foil facing, or with no facing material.

Aluminum Foil Insulation   Aluminum foil insulation is effective in controlling radiant heat, but to be effective it must have an air space between the foil and the wall surface. Aluminum foil is also an effective vapor barrier.

Loose Fill Insulation   Loose fill insulation is usually used to insulate ceilings in residential construction. This insulation is usually made from glass fiber, mineral wool, or expanded vermiculite. Glass fiber or mineral wool insulation may be placed by hand or by machine. It is usually placed by machine because machine application is more uniform

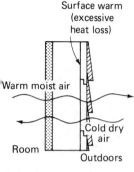

Surface warm
(excessive
heat loss)

Warm moist air

Cold dry
air

Room

Outdoors

No insulation — moist air
passes through the wall

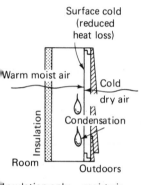

Surface cold
(reduced
heat loss)

Warm moist air

Cold
dry air

Condensation

Room

Outdoors

Insulation only — moist air
passes into wall and
condenses. Condensation
causes damage to insulation
and the building framework

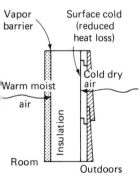

Vapor
barrier

Surface cold
(reduced
heat loss)

Cold dry
air

Warm moist
air

Room

Outdoors

Insulation with vapor barrier
— moisture cannot enter
wall space

Fig. 9-1 Controlling Moisture

and more economical. Vermiculite insulation is installed by manual methods.

CONTROLLING MOISTURE

Water vapor contained in air can readily pass through most building materials used for wall construction. This vapor caused no problem when walls were porous because it could pass from the warm wall to the outside of the building. When builders started to install insulation in the walls to cut down on heat loss, moisture in the air passed through the insulation until it reached a point cold enough to cause it to condense (see Fig. 9-1). The condensed moisture froze in very cold weather and reduced the efficiency of the insulation.

The ice contained within the wall thawed as the weather warmed, and the resulting water in the wall caused studs and sills to decay over a period of time.

Vapor Barriers

To prevent water vapor from entering insulated walls vapor barriers are installed on the warm side of the wall. The vapor barrier is a material through which moisture cannot penetrate very readily. Commonly used vapor barrier materials are asphalted kraft paper, aluminum foil, and polyethylene film.

Asphalted kraft paper is usually incorporated with blanket or batt-type insulation. It serves as a means for attaching the insulation to the building framework and as a vapor barrier when installed on the warm side of the wall.

Aluminum foil may be incorporated with blanket or batt-type insulation in the same manner as kraft paper. It is also applied to the back of gypsum lath and gypsum wallboard, where it works as an effective vapor barrier.

Polyethylene film is applied over the studs and ceiling joists after the insulation is installed. When polyethylene film or foil backing on wallboards is used the insulation will be plain batts or blankets which are held in place by friction. Polyethylene film as a vapor barrier is stapled over the studs and also covers the window frames. This aids in keeping the frames and sashes clean during application and finishing of the gypsum wallboard.

Fig. 9-2   Ventilating Gable and Hip Roofs (Courtesy Owens-Corning Fiberglas Corporation)

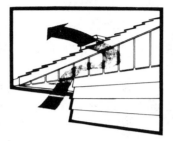

Gable vent–Use with eave vents or place at opposite gable ends for air flow.

Roof vent–Install on hip roofs with eave vents for best results.

Eave vent–Use with gable, ridge, or roof vents for air flow to remove moisture.

Roof stack–Use on flat roofs with wall vents to vent space above insulation.

Fig. 9-3 Ventilating Flat Roofs (Courtesy Owens-Corning Fiberglas Corporation)

Ventilation

Moisture in unheated areas such as attics or crawl spaces can be controlled by ventilating the area. The flow of air through these spaces carries the moist air to the outside before the moisture can condense on the framing members and sheathing. The amount of ventilation required varies with different climates and different building codes.

Attics of gable and hip roofs may be ventilated with a variety of louvers and vents. The use of ridge vents, gable louvers, roof vents, and eave vents is illustrated in Fig. 9-2. Flat roofs are ventilated with a combination of eave vents and roof stacks (see Fig. 9-3).

INSTALLING INSULATION

To be effective insulation must be properly installed. This is true whether the insulation is made of glass fiber, mineral wool, or some other material. The following discussion will cover the installation of blankets and batts in walls, ceilings, and floors.

Friction-Fit Wall Insulation

Friction-fit batts are available in 15″ and 23″ widths to fit standard 16″ and 24″ on center framing. The batts are placed between the studs and tightly fill the space between the studs, the sheathing, and the face of the studs. Small and irregular spaces are filled with pieces of batt insulation cut to fit the space. Very narrow spaces around door and window frames are fitted with small pieces of insulation which completely fill the space.

Because friction-fit batts do not have an attached vapor barrier, a separate vapor-resistant film (usually polyethylene film) must be applied over the studs (see Fig. 9–4). The usual procedure is to unroll the film across the entire wall, including door and window openings. The film is stapled to the top and bottom plates as it is unrolled and fastened securely to the studs. The vapor barrier must also be stapled to the framework around the perimeter of door and window openings before the openings are cut in the barrier. Often polyethylene film vapor barriers are left over the windows to act as a mask when the walls are spray painted. Laminated kraft paper vapor barriers are installed in the same manner as polyethylene film.

Faced Wall Insulation

Insulation with kraft paper or aluminum foil facings are available in widths for all standard stud spacings. They may be installed by inset stapling or by face stapling.

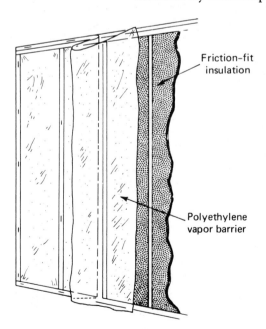

Friction-fit insulation

Polyethylene vapor barrier

Fig. 9-4   Polyethylene Vapor Barrier

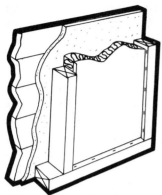

Inset stapling-Insulation is pushed into the stud or joist space and the flanges stapled to the sides of the studs or joists.

Fig. 9-5   Inset Stapling (Courtesy Owens-Corning Fiberglas Corporation)

When inset stapling is used the flanges are stapled snugly to the sides of the studs. The insulation at the top and bottom plates is peeled back about 1″ to form a flange for stapling (see Fig. 9-5).

If face stapling is used the flanges are stapled to the face of the studs and pulled tight. The insulation at the top and bottom plates is pulled back about 1″ to form a flange at the ends of the insulation. These flanges are stapled tightly to the plates (see Fig. 9-6). Face stapling provides a better vapor barrier than inset stapling but may cause some irregularities in gypsum wallboard if not applied carefully.

Care must be taken to install insulation tightly from bottom to top plates. Improperly installed insulation which does not reach the top and bottom plates allows convection currents to flow within the wall (see Fig. 9-7). These air currents pick up heat on the warm side of the wall and carry it to the cold side of the insulation. This flow of air reduces the efficiency of the insulation and can be avoided by following proper installation techniques.

Irregular stud spaces narrower than the regular spacings are insulated by cutting the insulation 1″ wider than the space between the studs. The flange on one edge of the insulation is stapled in the normal manner, and the flange on the other edge is formed by forcing the insulation between the studs and peeling the extra width of facing away from the insulation. The

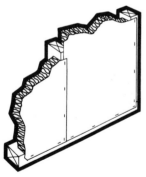

Face stapling-flanges are stapled to the edges of the studs or joists.

Fig. 9-6   Face Stapling (Courtesy Owens-Corning Fiberglas Corporation)

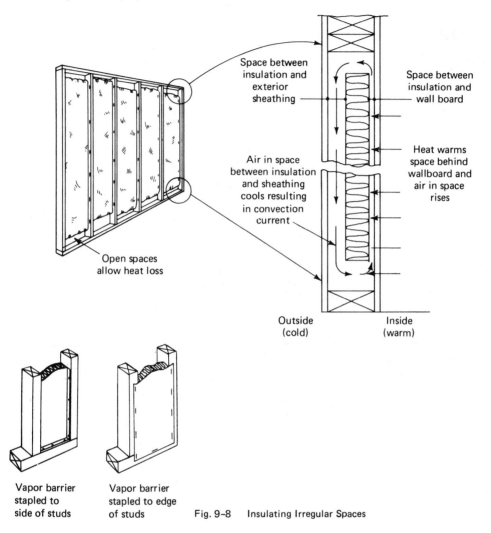

Fig. 9-7   Improper Installation

Space between insulation and exterior sheathing

Space between insulation and wall board

Heat warms space behind wallboard and air in space rises

Air in space between insulation and sheathing cools resulting in convection current

Open spaces allow heat loss

Outside (cold)

Inside (warm)

Vapor barrier stapled to side of studs

Vapor barrier stapled to edge of studs

Fig. 9-8   Insulating Irregular Spaces

flange formed is stapled to the stud in the usual manner (see Fig. 9-8).

Stud spaces wider than the available insulation are insulated horizontally. Pieces of insulation are cut approximately 2″ longer than the space between the studs, and a stapling flange is formed on each end. After the insulation is installed a separate vapor barrier is applied over the studs (see Fig. 9-9).

Small spaces between rough framing members and between door and window frames and rough framing may be filled with

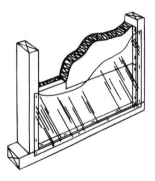

Strips of insulation
overlapped and covered
with a vapor barrier

Fig. 9-9   Insulating Wide Stud Spaces

hand-stuffed insulation. Narrow strips of vapor-resistant material should be stapled over all hand-stuffed areas (see Fig. 9-10).

When insulating around electrical outlets, as much insulation as possible should be pushed behind the outlet box. Insulation should be compressed and placed behind water pipes to prevent freezing in cold weather (see Fig. 9-11).

The placement of hot air heating and cooling ducts in outside walls should be avoided. However, when the ducts are

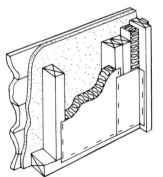

Insulation packed
into narrow space
and covered with
vapor barrier

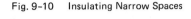

Fig. 9-10   Insulating Narrow Spaces

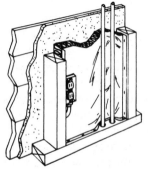

Insulation is forced
behind pipes and outlets

Fig. 9-11   Insulating around Pipes and Boxes

placed in outside walls insulation should be placed behind the ducts to avoid unnecessary heat loss.

Ceiling Insulation

Ceilings may be insulated with loose fill, friction-fit, or faced insulations. Loose-fill insulation is installed after the ceil-

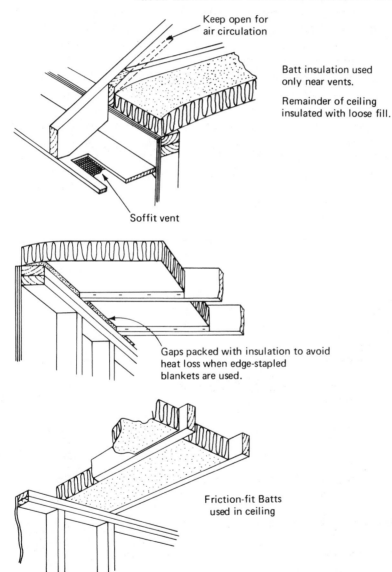

Keep open for air circulation

Batt insulation used only near vents.

Remainder of ceiling insulated with loose fill.

Soffit vent

Gaps packed with insulation to avoid heat loss when edge-stapled blankets are used.

Friction-fit Batts used in ceiling

Fig. 9–12 Insulating Ceiling with Friction Fit Insulation

ing material is in place. The vapor barrier, if used, must be applied to the ceiling joists before placing the ceiling material.

Friction-fit insulation is installed by placing it between the joists and allowing it to overlap the top plate of the outside wall (see Fig. 9–12). The insulation should touch the top plate along its entire width to prevent air infiltration and heat loss, but it should not completely block the ventilation space between the plate and roof sheathing if eave vents are used. Laminated kraft paper or 4-mil polyethylene film should be stapled to the ceiling joists to control moisture and to help control humidity levels in the building.

Faced insulation may be installed by inset stapling or face stapling to the ceiling joists. Inset stapling is required to get maximum insulation value from foil faced insulation.

When inset stapling is used the insulation should be pulled down snugly to the plate at the outside wall and should have sufficient overlap at the wall to prevent infiltration. The excess facing of the insulation should be stapled to the top plate. A similar procedure is followed when face stapling is used (see Fig. 9–13).

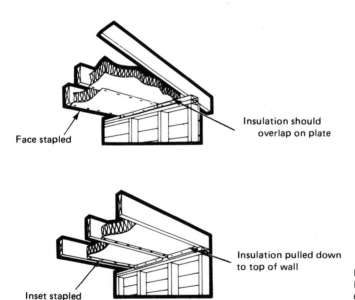

Face stapled

Insulation should
overlap on plate

Insulation pulled down
to top of wall

Inset stapled

Fig. 9–13   Insulating Ceilings with Faced Insulation (Courtesy Owens-Corning Fiberglas Corporation)

Floor Insulation

Floor insulation is installed with the vapor barrier toward the floor or warm-in-winter side of the insulation. Reverse flange insulation is often used for this purpose. The nailing flange is attached to the special breather back paper in a manner which allows conventional inset stapling (see Fig. 9–14).

Other methods of installing floor insulation may be used, but most of these require some type of wire or wire screening to hold the insulation in place.

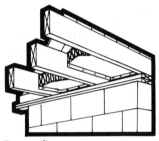

Reverse flange.

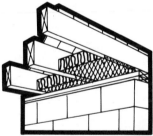

Insulation supported on chicken wire.

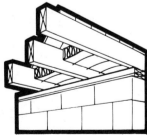

Insulation supported on heavy wire.

Fig. 9–14   Insulating Floors (Courtesy Owens-Corning Fiberglas Corporation)

REVIEW QUESTIONS

1. What are the functions of insulations?
2. How does temperature change affect comfort?
3. Name some of the various types of insulation.
4. Where is blanket-type insulation used?
5. Where is batt-type insulation used?
6. Where is loose-fill insulation used?
7. How is water vapor controlled when insulating?
8. How is friction-fit insulation installed?
9. How is faced wall insulation installed?
10. How are ceilings insulated?

interior
wall materials

10

Interior walls and ceilings in residential construction are finished with a variety of different materials. Among the most commonly used groups of materials are lath and plaster, thin-coat plaster, gypsum drywall, prefinished paneling, and solid wood paneling. Not all of these materials are installed by carpenters, but the carpenter must work around all of them and should be aware of their characteristics.

LATH AND PLASTER

When conventional lath and plaster is used the carpenter must install ground strips, which are actually guides for the plasterer. Ground strips are usually ¾″ square but vary in size with the finished thickness of the lath and plaster and other job conditions. They are commonly installed around the perimeter of the room at the floor line, around all interior door openings (see Fig. 10-1), and in other areas where a guide is needed.

In buildings where very narrow casings are used around

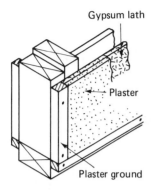

Gypsum lath

Plaster

Plaster ground

Fig. 10-1 Typical Plaster Grounds

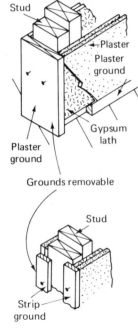

Stud

Plaster

Plaster ground

Gypsum lath

Plaster ground

Grounds removable

Stud

Strip ground

Fig. 10-2 Removable Grounds

doors it is necessary to use a ground which can be removed after the plastering is completed. This ground is as wide as the full thickness of the finished wall and is temporarily nailed around the perimeter of the opening (see Fig. 10-2).

In most cases it is not necessary to place grounds around exterior doors or windows, because the jambs and sill of the frames project beyond the studs a distance equal to the thickness of the lath and plaster. The jambs and sill of the frames then act as the ground or guide for the plasterer. When grounds are needed around window frames they are installed in the same manner as around interior door openings.

Plaster Bases

Nearly all plaster in residential work is applied over gypsum lath. This lath is usually nailed directly to the building framework, but for added soundproofing and protection against cracks it may be attached with resilient clips (see Fig. 10-3).

Plain gypsum lath is available in 3⁄8″ and 1⁄2″ thicknesses. Gypsum lath 3⁄8″ thick is used when stud and joist spacing is 16″ or less, and 1⁄2″-thick lath is used when joists and studs are spaced over 16″ but not more than 24″ on center. The standard size for gypsum lath of both thicknesses is 16″ by 48″. In some areas it is available in 8′ and 12′ lengths.

Insulating gypsum lath is plain gypsum lath with aluminum foil laminated to the back face. The foil provides a vapor barrier and insulates for radiant heat when it faces an air space.

Metal lath is seldom used in residential work except to meet special conditions and to provide reinforcing over gypsum lath at inside corners and other areas where cracking may occur.

Plastering Accessories

Various accessories in the form of corner beads, casing beads, and screeds are available for use with lath and plaster. These are installed over the lath and provide a guide for the plasterer. All outside corners require corner beads for reinforcing. These beads also provide a guide for the plasterer and guarantee a straight line at the corners. Screeds are seldom used in residential work, but when they are used on long walls they provide an additional guide for the plasterer.

Fig. 10-3 Gypsum Lath Attached with Clips (Courtesy National Gypsum Company)

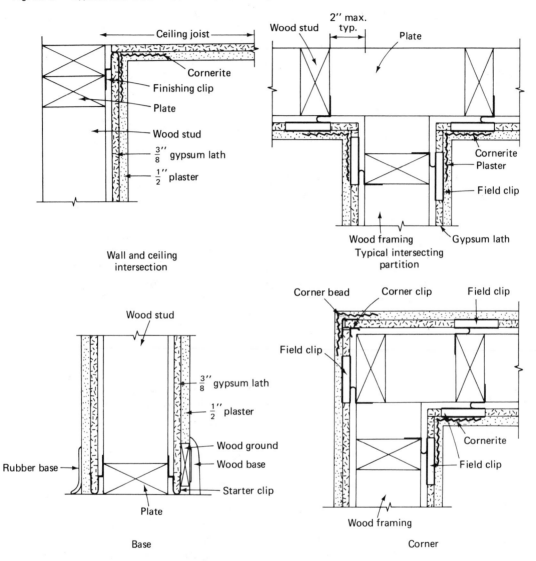

Wall and ceiling intersection

Typical intersecting partition

Base

Corner

Plastering

Two coats of plaster are applied over gypsum lath. The first coat fills most of the space between the lath and the ground strip. It is usually a mixture of gypsum plaster, an aggregate such as sand or perlite, and water. Various base-coat plasters are available for use over gypsum and metal laths.

The second coat of plaster applied in two-coat work is known as the finish coat. It fills the irregularities in the base coat and gives the wall its finished surface. Most plaster walls are given a sand float finish, but kitchens, bathrooms, and areas to be wallpapered are often given a smooth troweled finish.

The lath-and-plaster system of walls and ceilings introduces a quantity of water into the building framework. This water must be removed from the building by ventilating the building. Heating is required in cold weather to prevent freezing and to aid the drying process. The time required to dry the walls and building framework varies with weather conditions but averages about one week.

Work on finish floors and trim cannot begin until the drying period is over. This delay in completing the building is one of the major reasons that the use of lath and plaster in residential construction has declined. However, the major gypsum manufacturing companies have overcome the problem of moisture being introduced into the building with the development of thin-coat plastering systems.

THIN-COAT PLASTER

Thin-coat plaster, or veneer plaster as it is often called, employs specially formulated plaster applied over a special large-size plaster base. Veneer plasters are applied in one- or two-coat operations to a thickness of $1/16''$ to $3/32''$ per coat.

Veneer Plaster Base

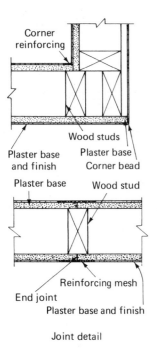

Corner reinforcing

Wood studs

Plaster base and finish

Plaster base

Corner bead

Plaster base

Wood stud

Reinforcing mesh

End joint

Plaster base and finish

Joint detail

Fig. 10–4   Thin Coat Plaster Details

Veneer plaster bases are available in thicknesses of $1/2''$ and $5/8''$, a width of $4'$, and lengths from $8'$ to $16'$. The $1/2''$ material is for application over framing members $16''$ on center, and the $5/8''$ material is used when the framing members are $24''$ on center, or when extra strength and fire resistance are required.

The large-size sheets are nailed to the building framework in such a manner as to give a minimum number of end joints. All end and edge joints are reinforced with a $2\frac{1}{2}''$-wide woven fiberglass tape. This tape is securely stapled to the plaster base at $12''$ to $24''$ intervals. Inside corners are reinforced with the same tape, which is folded into the corner (see Fig. 10-4).

As with regular plaster systems, outside corners of veneer plaster systems are reinforced with metal corner beads. Metal beads are also available for casings and expansion joints.

Veneer Plastering

Each manufacturer of veneer plastering products makes specific recommendations for the mixing and application of the veneer plaster, which can be applied in one-coat or two-coat operations. Veneer plasters can be troweled smooth, or they can be floated or textured in the same manner as regular plaster.

The veneer plasters offer a hard surface, fast installation, ease in decoration, and effective sound control. In many cases veneer plaster can be installed and finished more quickly than gypsum drywall.

GYPSUM DRYWALL

Gypsum wallboard is a mill-fabricated product composed of a fireproof gypsum core with a smooth heavy paper on the face side and a strong liner paper on the back side. The face paper on all gypsum wallboard is folded around the long edges to reinforce the board. The ends of the sheets are cut smooth and square.

Gypsum drywall is presently used on more walls and ceilings in residential construction than any other material. Because it is a type of wallboard it is commonly installed by carpenters. In many areas carpenters who install gypsum wallboard have become specialists and are called drywall applicators.

Drywall used in residential construction is usually ½″ thick, but ⅝″-thick drywall is available for added resistance to fire and sound transmission. Gypsum wallboard is also manufactured in ⅜″ and ¼″ thicknesses for use in covering old walls and other special purposes. The standard width for gypsum wallboard is 4′. It is manufactured in lengths from 8′ to 16′.

Drywall Installation

Gypsum drywall is usually installed in a single layer applied by nailing directly to the studs and joists. For added resistance to nail pops, caused by loose nailing and lumber shrinkage, an adhesive may be applied to the framework before the drywall is nailed in place. The adhesive takes up small irregularities and holds the drywall panel securely. Special screws may also be used to fasten the drywall securely to the studs and joists.

A double-layer system made up of two layers of ⅜″ drywall is sometimes used. The joints of the two layers are staggered to provide adequate joint reinforcing. The layers are nailed to the framework and bonded together with a special adhesive.

Single-Layer Application   In single-layer application in residential construction, with few exceptions, it is usually more desirable to apply the gypsum drywall at right angles to the framing. The layout should be planned to minimize the number of end joints. All end joints should occur at a framing member, and away from the center of the room whenever possible. The number of end joints can be limited by using the longest possible length of wallboard which can be handled conveniently. End joints in adjacent rows of drywall should be staggered and made at opposite ends of the room whenever possible.

Normally drywall is applied to the ceiling first. The sheets are cut to length as needed by scoring the face paper with a wallboard knife, snapping the core, and then cutting the back paper. Adjacent sheets should be fitted closely together, but they should not be forced into place.

Using the double-nailing method, the wallboard should first be held tight against the ceiling joists and nails driven every 12″ along the support. A second nail should be driven 2″ to 2½″ from the first to draw the board to the joists. Single nails are placed at the edges and ends of the sheet. Nails at the ends of the sheet should be spaced not more than 7″ on center.

Wallboard on side walls is held tight against the studs and nailed using the double-nailing method as in the ceiling. The double-nailing method is illustrated in Fig. 10–5. On side walls the nails at end joints may be spaced 8″ on center. Care must be taken to drive nails without tearing the face paper. Nails must be driven below the surface of the board with the head of the hammer leaving a dimple (see Fig. 10–6). This depression or dimple accepts joint and finishing compound, which is used to conceal the nail heads.

Nails at interior angles where walls or walls and ceilings meet may be omitted to form a floating interior angle (see Fig.

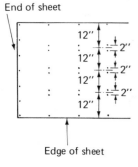

Fig. 10–5   Double Nailing for Gypsum Wallboard

Improper nailing

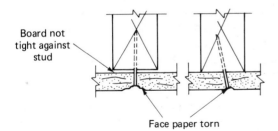

Board not
tight against
stud

Face paper torn

Proper nailing

Board tight
against stud

Face paper "dimpled"   Fig. 10-6   Nailing Gypsum Wallboard

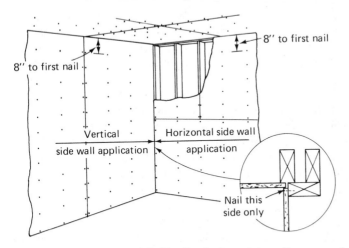

8" to first nail

8" to first nail

Vertical
side wall application

Horizontal side wall
application

Nail this
side only

Fig. 10-7   Gypsum Wallboard Installed with Floating Interior Angles

10-7). Omitting the nails near the angle as illustrated reduces cracking and nail pops caused by movement in the building framework.

Openings in the walls and ceilings for electrical outlets may be made with a special die-cutting tool or with a keyhole saw. The wallboard knife may also be used to cut openings. When this is done, the face paper is scored to conform to the size of the opening and the core is broken out with a hammer and pulled clean from the back side of the wallboard.

Double-Layer Application When a double layer of wallboard is used a face layer of gypsum wallboard is applied over a base layer of gypsum or fiber wallboard which is fastened to the building framework. The joints between the layers are offset at least 10″ to provide a good bond between layers and to prevent cracking.

The base layer may be fastened to the framework with screws, nails, or staples. Drywall screws should be placed every 16″ when framing members are 16″ on center and every 12″ when the framing members are 24″ on center.

When the double-nailing system is used on the base layers, double nails are placed every 12″ as recommended for single-layer application (see Fig. 10-5). Single nails and staples should be placed every 7″ on the ceiling and every 8″ on side walls. Care should be taken to avoid cutting the face paper with the nail or staple heads.

The face layer is applied with an adhesive recommended by the wallboard manufacturer and held in place with nails or bracing until the adhesive bonds the two layers together.

Finishing Joints and Nails After the gypsum wallboard has been fastened to the building framework all joints and nail heads must be concealed and finished. All edge joints are reinforced with a special paper tape imbedded in joint compound. The tape and the joint compound are applied in the recess formed by the tapered edge of the wallboard and form a flat smooth surface between the sheets of wallboard (see Fig. 10-8).

End joints are more difficult to finish because they are not provided with a taper or recess for the tape and compound. The tape is imbedded in the initial coat of compound which has been applied over the joint and is immediately covered with compound to prevent loss of bond at the edges of the tape, which causes curling.

After the first coat is dry a second and third coat are applied. Each successive coat is feathered out wider than the previous coat in an effort to make the joint as nearly invisible as possible.

In may be necessary to sand the joints between coats, depending on the skill of the finisher and the desired finished appearance.

Nail heads are "spotted" with joint compound three or

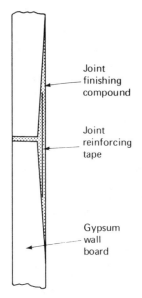

Joint finishing compound

Joint reinforcing tape

Gypsum wall board

Fig. 10-8 Cross Section Finished Tapered Edge Joint

four times, depending on the depth of the fill, drying conditions, and desired finish. Light sanding may be necessary after the final coat for hard-to-remove small irregularities.

Various texture paint products are available for finishing gypsum wallboard. Some may be applied with a brush, roller, or sprayer, but others are made for sprayer application only. These materials are usually applied by professional painters who are skilled in the application of drywall texture paints.

PREFINISHED PANELING

Prefinished paneling is made from a number of materials, but the most commonly used panels are made of either hardboard or plywood. Hardboard is manufactured with a number of wood grain printed surfaces which simulate natural wood, and it is also available in marble patterns and solid colors. Because the hardboard used to make prefinished paneling is comparatively hard it is resistant to more abuse than some other types of paneling.

Prefinished plywood paneling is manufactured from nearly every commercial hardwood. The cost of the panel varies with the kind of wood. Most prefinished panels have randomly spaced V grooves. Even though the grooves are randomly spaced there is a groove every 16" across the sheet. This spacing makes it possible to conceal nails in the grooves.

Nearly all types of prefinished panels may be fastened to the building framework with adhesives, nails, staples, or screws. Adhesives offer the advantage of avoiding holes in the paneling which must be filled, but some method must be found to hold the panels in place until the adhesive sets.

Installing Prefinished Paneling

Before attempting to install any paneling the framing, furring, or solid backing to which the panel will be fastened should be checked for alignment. Any necessary adjustments for humps or hollows in the framework are made before the first panel is fitted and fastened in place.

Studding which is out of line may be shimmed with

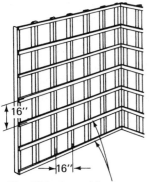

1" x 2" horizontal furring
strips — shimmed as necessary
to obtain flat and true plane

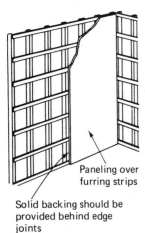

Paneling over
furring strips

Solid backing should be
provided behind edge
joints

Fig. 10-9   Furring Strips for
Paneling Applied over Wood
Framing

wedges cut to size, or furring strips may be nailed on the side of the stud to produce a straight nailing surface.

Furring strips are installed over concrete or masonry walls to provide a flat surface on which the paneling can be fastened with nails or adhesive. These strips are installed vertically 16" on center so that there is solid support at each joint between panels. If it is necessary to install the furring horizontally the 16" on center spacing is maintained. In addition, vertical furring must be installed every 48" to provide backing at the joints between panels.

Vertical or horizontal furring strips on existing frame walls may be fastened by nailing directly into the studding. Various adhesives are also used to facilitate the application of furring strips over existing walls (see Fig. 10-9).

Furring strips may be fastened to masonry walls with hardened nails, screws and masonry anchors, or various adhesives (see Fig. 10-10). Care must be taken to keep the face of all furring strips in one plane, because the paneled wall will be only as straight as the furring to which it is applied.

Fastening panels to the wall may be accomplished by nailing, panel adhesives or mastic adhesives. When nails are used they should penetrate at least ¾" into the studs or furring (see Fig. 10-11). Special hardened nails colored to match the paneling are available. When used these nails are driven flush to the surface of the paneling. Setting the nail head below the surface of the panel is not required.

Finishing nails are usually placed in the grooves of the panels and set below the surface. At the edges of the panel, where it is not possible to place nails in the groove, the nails are placed approximately ½" from the edge and are set below the panel surface. The nail heads are concealed by application of a putty colored to match the paneling. Most panel manufacturers also manufacture a putty stick colored to match the various panels.

When painted nails which match the color of the paneling are used it is not necessary to set the head of the nail below the panel surface. The small flat heads of these nails blend into the panel and are inconspicuous. Caution must be used with colored panel nails as they are usually hardened and will break easily. Because careless driving may cause them to break and fly, there is danger of eye injuries. Therefore always wear safety glasses when driving hardened nails.

Various panel adhesives are available for installing prefinished panels. Most of these are packaged in cartridges for

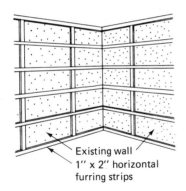

Existing wall
1″ x 2″ horizontal
furring strips

Horizontal furring

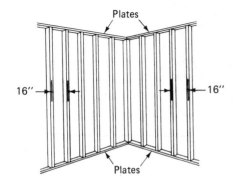

Plates

16″   16″

Plates

Vertical furring — 1″ x 2″,
2″ x 2″, or 2″ x 4″ may be
used

Panel

Furring on masonry with adhesive panel

Note: when applying
furring to masonry
walls with adhesive
use a continuous zig-
zag bead. Use the
adhesive as a shim
by filling voids
between furring and
irregular masonry.
Wait at least 24
hours before panel
application

Fig. 10-10   Installing Furring Strips on
Masonry Walls

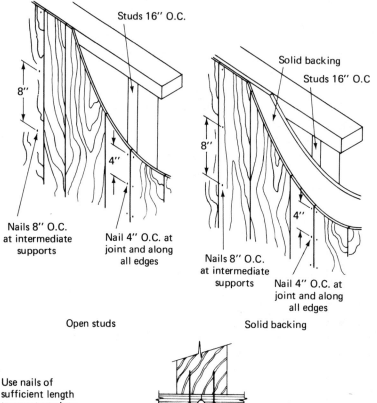

Studs 16" O.C.

Solid backing

Studs 16" O.C

8"

8"

4"

4"

Nails 8" O.C.
at intermediate
supports

Nail 4" O.C. at
joint and along
all edges

Nails 8" O.C.
at intermediate
supports

Nail 4" O.C. at
joint and along
all edges

Open studs

Solid backing

Use nails of
sufficient length
to penetrate into
studs at least $\frac{3''}{4}$

Note: matching groove at joint

Fig. 10–11 Panel Fastened
with Nails

application with a caulking gun. Because of the many different
types of adhesives available it is recommended that specific
manufacturers' instructions be followed.

All panel adhesives require clean and dry surfaces. In general they are applied in a continuous bead along the center of
the studding or furring, but for some types of adhesives an
intermittent bead 3" long with a 6" space is acceptable on intermediate studs (see Fig. 10–12).

After the adhesive is applied the panels are set into place
and pressed firmly to the framework to attain an initial bond
with the adhesive. A minimum of two nails are placed at each
end of the sheet to hold it in place while the adhesive cures.
After a short time additional pressure is applied to all adhesive
areas to obtain a final bond.

Contact cement is sometimes used to bond paneling to

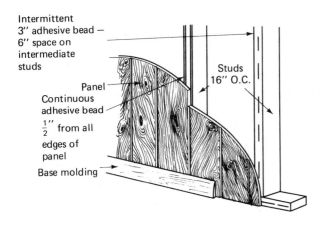

Intermittent 3" adhesive bead — 6" space on intermediate studs

Studs 16" O.C.

Panel

Continuous adhesive bead $\frac{1}{2}$" from all edges of panel

Base molding

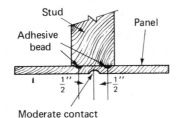

Stud

Panel

Adhesive bead

$\frac{1}{2}$" $\frac{1}{2}$"

Moderate contact

Fig. 10-12   Panel Fastened with Panel Adhesive

Existing backing (wood, plaster, gypsum board)

Base molding

Waterproof adhesive Mastic type

Fig. 10-13   Panel Fastened with Mastic Adhesive

existing walls or directly to studs or furring. When contact cement is used all surfaces must be clean and dry. A coat of cement is applied with a brush or roller to the surfaces which will be bonded together. After the cement on both surfaces is dry the panel is positioned carefully and set into place. It bonds on contact with the prepared surface and cannot be moved. Therefore extreme care must be taken in positioning the panel.

Mastic cements are used to fasten prefinished panels over existing walls of plaster, wallboard, or wood (see Fig. 10-13). The adhesive is applied over the entire area, which must be clean and dry. Panels are positioned and pressed against the mastic. The mastic will generally hold the panels in position without nails. However, if there are any unusual conditions a small nail can be placed near the corner of the panel to hold it in position until the mastic sets.

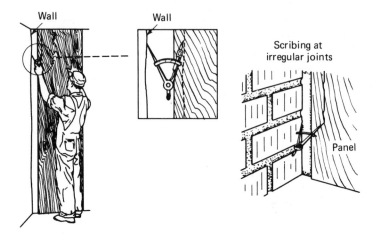

Fig. 10–14   Scribing a Panel

Panel Installation Procedure

The first panel to be installed is usually placed at an inside corner and held in a plumb position. If necessary, it is scribed to the adjacent wall (see Fig. 10-14). After scribing, the panel is placed on a pair of saw horses, cut with a fine-tooth saw, and trimmed with a block plane as necessary. When the necessary cutting is completed the panel is put in place and checked for plumb with a carpenter's level. If ceiling and base trim will be used, no further fitting is necessary, and the panel may be fastened in place.

Sometimes no moldings are used at the ceiling line. Then it is also necessary to scribe the panel to the ceiling. This is done after the panel is scribed and fitted to the first corner. It is placed into the corner and raised to the ceiling while the edge is maintained in a plumb position. The top edge is scribed to the ceiling, and it is cut and fitted in the same manner as the first edge. When the trimming and fitting is completed the panel is placed in position, checked for alignment, and fastened to the wall.

Succeeding panels are cut to length, if necessary, and held in position along previously placed panels. The top end is scribed and fitted to the ceiling as necessary, and the panels are fastened in place by any of the previously described methods.

On new work, where door and window trim has not yet been installed, the paneling is cut in a manner which will allow it to fit up to the frame (see Fig. 10-15). Only normal care is

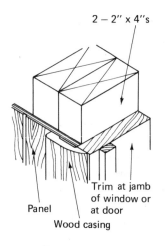

Fig. 10-15   Fitting Panels at Doors and Windows

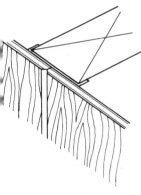

A. Standard joint

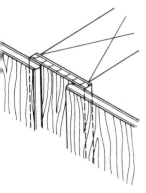

B. Wipe groove

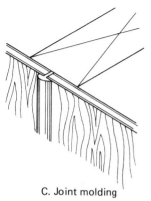

C. Joint molding

Fig. 10–16   Joints between Panels

required when making this cut because the finish trim around the door or window openings will cover the edge of the panel.

Base trim is used to cover the joint between the panel and the floor. This base trim is installed in the same manner as baseboard is installed over other wall materials (see Chapter Twelve).

Joints Between Panels   Most prefinished panels are joined by simple butt joints which are made inconspicuous by a V groove (see Fig. 10–16A). A wide joint is sometimes used and may be obtained by placing a strip of prefinished paneling below the adjoining panels (see Fig. 10–16B). When this system is used the entire wall must be furred out to provide necessary backing for the large face panels.

A special joint molding (see Fig. 10–16C) is sometimes used. This molding is placed over the edge of the panel after the panel is installed. The molding is then fastened to the wall framework by nailing through its exposed leg.

Corner Joints   Inside corner joints are usually scribed and butted. However, special inside corner moldings are available for use with many different types of panels (see Fig. 10–17). Many of these moldings are finished to match the type of paneling being installed.

Outside corners of prefinished paneling may be mitered as shown in Fig. 10–17. The mitered corner gives a pleasing appearance. Unfortunately it can withstand only limited abuse and should be used only where it is not subject to heavy traffic. In making the mitered corner a straight edge should be clamped in place to the back of the panel to provide a guide for the portable power saw. This procedure assures a straight cut and a well-fitting mitered corner.

Metal and wood outside corners are available for various types of prefinished panels. The metal type is installed before the second corner panel is put in place (see Fig. 10–17). Wood corner beads prefinished to match the paneling are applied after the panels are in place. Each type of molding offers a pleasing appearance and provides a durable corner which can withstand more abuse than the mitered corner.

When prefinished paneling is used on existing walls in remodeling work the paneling may be fitted to the window and door castings as shown in Fig. 10–18. In some cases the metal edge molding may be omitted and the paneling fitted closely to the door casing. The use of the molding makes fitting easier.

At inside corners

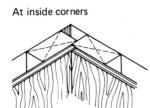

Butt joint
one panel scribed, other butted

At outside corners

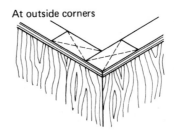

Mitered corner

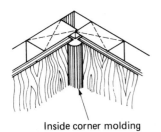

Inside corner molding

Molding

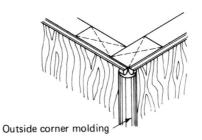

Outside corner molding

Molding

Fig. 10-17   Corner Finishing

Cut panels carefully to fit
around casing, making sure
that correct measurements
are transferred from wall
to panel

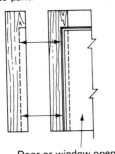

Door or window opening

2 – 2" x 4"s

Existing wall

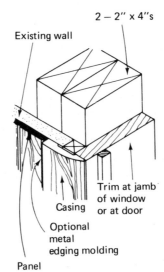

Panel

Casing

Optional
metal
edging molding

Trim at jamb
of window
or at door

Fig. 10-18   Fitting Panels to Existing Trim

SOLID WOOD PANELING

Solid wood paneling can be made from any kind of hardwood or softwood lumber. However, most solid wood paneling is manufactured from western pine lumber. It may be manufactured from lumber which has few if any defects as well as from less expensive grades of lumber. Some pine lumber is also selected for its quality of many tight knots and is manufactured into knotty pine paneling.

Many standard patterns of solid wood paneling are available, and paneling can be made in special patterns if required. Although there are many patterns of solid wood paneling available, local lumber dealers usually stock only three or four of the most popular patterns. Some of the patterns of paneling which are most commonly stocked are illustrated in Fig. 10-19.

Interior wood paneling is usually installed vertically, but it can also be installed horizontally or diagonally. It may also be installed in a combination of directions to obtain special effects (see Fig. 10-20).

When wood paneling is delivered to the job it should be stored for several days in the room where it will be installed. Separating sticks should be placed between the individual boards to allow air to circulate around them. The circulation of air allows the boards to dry out and adjust to the moisture conditions of the room. The added time taken to dry the panel-

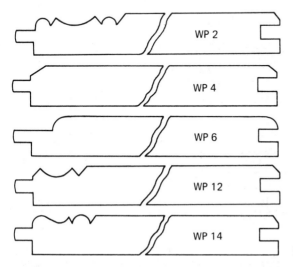

WP 2

WP 4

WP 6

WP 12

WP 14

Fig. 10-19   Paneling Patterns

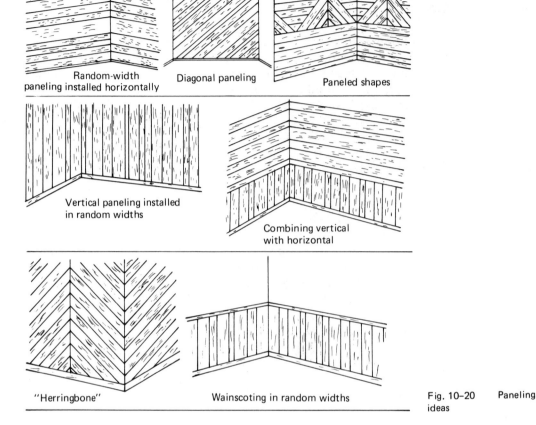

Random-width paneling installed horizontally

Diagonal paneling

Paneled shapes

Vertical paneling installed in random widths

Combining vertical with horizontal

"Herringbone"

Wainscoting in random widths

Fig. 10–20   Paneling ideas

ing to room conditions will pay off in neatly fitting boards which will not shrink or swell after being installed.

Installing Wood Paneling

Before installing wood paneling the wall must be prepared by installing furring and backing. This is done in a manner similar to that for furring supporting prefinished paneling, the main difference being that less furring is needed because the solid paneling is usually ¾" or $^{25}/_{32}$" thick.

When solid paneling is applied horizontally over wood studding no additional furring is needed, since the paneling can be nailed directly to the studs. On concrete or masonry walls it

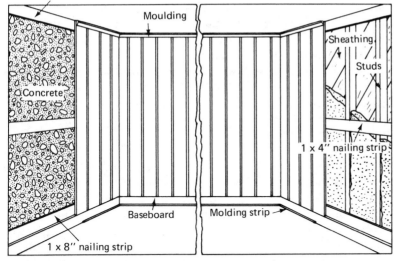

1 x 4" nailing strip

Moulding

Sheathing

Studs

Concrete

1 x 4" nailing strip

Baseboard

Molding strip

1 x 8" nailing strip

Basement   Wood frame   Construction

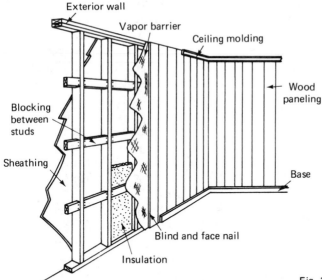

Exterior wall

Vapor barrier

Ceiling molding

Wood paneling

Blocking between studs

Sheathing

Base

Blind and face nail

Insulation

Fig. 10–21   Furring for Wood Paneling

is necessary to install furring on 16" to 24" centers to provide a nailing surface.

When the paneling is installed vertically on wood frame walls it is necessary to apply 1 by 4 furring strips at right angles to the studs to provide backing for the paneling. It is usually sufficient to install this furring at the base line, midway be-

All patterns of solid wood paneling are blind nailed

Fig. 10-22   Nailing Wood Paneling

Gypsum drywall ceiling

Molding

Frieze

Careful fit required

Paneling

Stud

Careful fit required at this point

Furring

Base

Finish floor

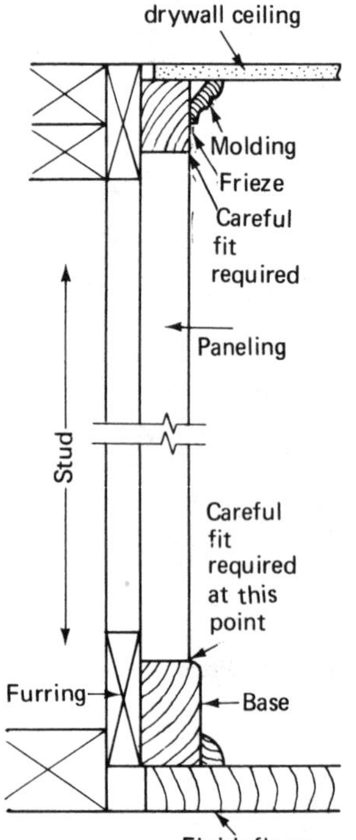

Fig. 10-23   Paneling Fitted between Base and Frieze

tween the floor and ceiling, and at the ceiling line (see Fig. 10-21). In no case should the furring spacing exceed 48″.

If the paneling is installed in a combination of directions the furring must be placed in a manner which will provide the necessary backing for the paneling.

Wood paneling is fastened in place by blind nailing whenever possible (see Fig. 10-22). Generally the only face nails which are applied are in the first and last panel boards. These nails are made comparatively inconspicuous by setting them below the surface of the panel.

Finish nails 2″ long (6d) are usually used to fasten solid wood paneling. In some cases it may be desirable to use smaller (4d) nails, but at other times longer nails (8d) may be needed.

Before starting the paneling job all furring must be installed and all necessary layout work completed. When boards of various widths are mixed to give the wall a more pleasing appearance it is most important to make a preliminary layout to avoid the need for installing a narrow strip of paneling at the end of the wall.

To make the layout for random placing of boards of different widths, short pieces of each width board are used to mark board location on the furring strips. By careful trial and error mixing of the various widths, it is possible to get the right combination so that both the starting and finishing boards will be nearly full width. Care must be taken to avoid ending with a molded edge in the corner as this condition gives a poor appearance.

Boards meeting at an inside corner are scribed in the same manner as prefinished paneling to obtain a good fit at the corner. Boards meeting at outside corners may be mitered to provide a finished corner, or a corner molding may be applied to cover the edges of the boards.

The joint between the paneling and the ceiling is usually concealed by a molding. The shape of the molding used will vary with the desired finished appearance, and therefore almost any pattern of molding may be used.

In some cases the paneling may run between the base and the frieze (see Fig. 10-23). When this is done the base should be installed level, and the frieze should be parallel to the base to facilitate the fitting of the panel boards. Great care must be taken when fitting boards between the base and frieze, since a small error in cutting to length can lead to a poor fit between the base and the panel or between the panel and the frieze.

If the paneling is set on the base but covered by a molding

at the ceiling (see Fig. 10-21) the job of installation is easier because only one end requires fitting.

REVIEW QUESTIONS

1. What are some of the materials used on interior walls?
2. What are ground strips?
3. What are plaster bases?
4. What are plaster accessories?
5. How is thin-coat plaster applied?
6. What is gypsum drywall?
7. What are some advantages of gypsum drywall?
8. How is gypsum drywall installed?
9. Describe the features of prefinished paneling.
10. How are prefinished panels installed?
11. How is solid wood paneling installed?

finish floors

11

The most commonly used finish floors in residential construction are made of wood. Wood floors are popular in residences because they offer a wide range of desirable characteristics and qualities. Among these are a variety of distinctive and attractive appearances to meet the desired decorative look. They provide a good degree of hardness and wearing qualities necessary for long service, but they also give a necessary degree of resilience that makes them comfortable to walk on. The low heat conductivity of wood gives it a feeling of warmth.

The various types of wood floors are easy to maintain and comparatively easy to install. They may be divided into three broad categories: strip flooring, parquet flooring, and underlayment for resilient flooring materials and carpeting.

STRIP FLOORING

Strip flooring is the most widely used finish flooring and generally the most economical. It is available in a number of standard patterns and is made of various species of hardwood

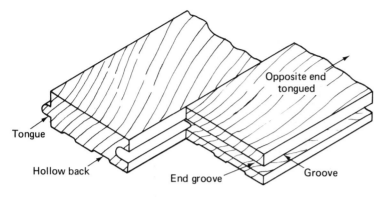

Tongue

Hollow back

End groove

Opposite end
tongued

Groove

Fig. 11-1   End Matched Strip
Flooring

and softwood. The most commonly available softwood flooring
is made from Douglas fir and southern yellow pine. Most hardwood strip flooring is made from oak and maple, but small
amounts of hardwood strip flooring are made from pecan,
hickory, ash, walnut, and other hardwoods.

Most strip flooring is tongue and groove, and all hardwood
strip flooring is also end matched (see Fig. 11-1). The hollow
back of hardwood strip flooring allows the boards to bridge
small irregularities and lay flat on the subfloor.

The building should be reasonably free of moisture before
any finish material is delivered. This means that there should be
adequate heat and ventilation in the building to dry the plaster
and the building framework. After the finish flooring is delivered it should be allowed to season for a time to reach an
equilibrium moisture content with that in the building. This is
of greatest importance when the flooring has been stored in a
damp area before delivery.

Installing Strip Flooring

Before installing strip flooring over any subfloor, the subfloor should be swept clean, and any foreign materials such as
mud, plaster, drywall joint cement, and so on should be scraped
from the floor. If the strip floor is being placed over a board
subfloor it is common practice to cover the subfloor with deadening felt or asphalt-saturated felt paper (see Fig. 11-2). This
paper prevents dust from working up through the floor, provides a sound deadener, and reduces air and moisture infiltration. The main purpose of felt paper on plywood subfloors is to
provide a sound deadener.

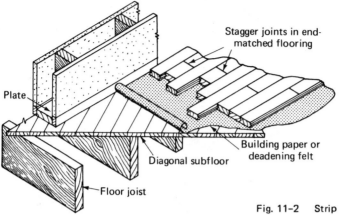

Fig. 11-2   Strip Flooring Applied over Deadening Felt

When the board subfloor is laid at a 45° angle to the joists the strip flooring can be laid at right angles to the joists. This is preferred because the flooring is given maximum support and a stiffer floor results.

If the subfloor runs at right angles to the joists the finish strip flooring must run at right angles to the subfloor. This makes the finish flooring parallel to the joists, and the resulting floor is not as strong as when the finish boards run across the joists. It is not possible to place the finish floor boards at right angles to the joists when the subfloor boards are placed in that direction because shrinkage of the subflooring would cause movement in the finish flooring, and the resulting spaces between finish floor boards is unacceptable.

Strip flooring can be installed in either direction over plywood subfloors because plywood is dimensionally stable (see Fig. 11-3) and there is no movement to cause spaces to open between the finish floor boards.

After the felt paper is in place the location of the joists should be marked on the paper with chalk lines to aid in nailing the boards in place. Whenever possible the finish floor should be nailed to the joists, because the nails will hold better in the joists than in the subfloor alone.

Racking Strip Flooring   Hardwood strip flooring is packaged in bundles of random-length pieces. Individual bundles will contain pieces nearly all the same length. The shortest-length bundles contain boards 2' long and under, while the longest bundles may contain boards up to 16' long.

Long boards should be placed near the wall to make starting along a straight line easier. Also, in many cases the boards

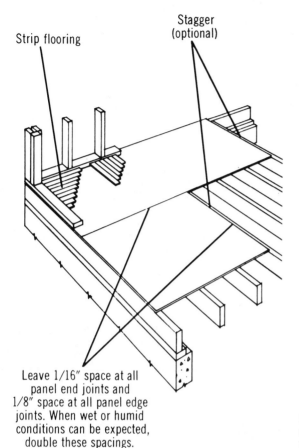

Strip flooring

Stagger
(optional)

Leave 1/16″ space at all
panel end joints and
1/8″ space at all panel edge
joints. When wet or humid
conditions can be expected,
double these spacings.

Fig. 11-3   Strip Flooring Installed over Plywood Subfloor (Courtesy American Plywood
Association)

near the wall will be exposed while boards in the center of the
room will be covered by rugs and furniture. Placing long boards
near the wall will provide the best possible appearance.

The boards are racked out or laid in position about 6″
from the wall or starting line in the direction parallel to the
starting line. The end-matched ends are laid end to end and by
sorting the various lengths of boards from two or more bundles
of flooring it should be possible to find pieces of the proper
length needed to finish the row of boards with little or no waste
(see Fig. 11-4). Joints between boards should be staggered at
least 6″ to provide a good appearance and to avoid weak spots
in the floor.

Nailing End-Matched Flooring in Place   After racking out a
number of rows of boards, sometimes for the full width of the

Fig. 11-4   Racking   End   Matched   Flooring
(Courtesy   Rockwell   Manufacturing   Company)

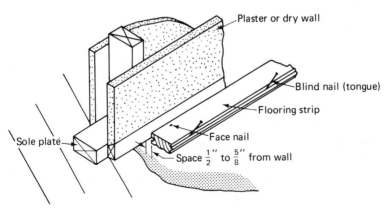

Fig. 11-5   Starting Strip
Floor Installation

room, the floor layer will start to nail the boards in place. Tongue-and-groove boards are blind nailed, except for the first row which is face nailed. The last few rows will also be face nailed because it is impossible to blind nail near the wall.

The starting row of strip flooring is installed ½″ to ⅝″ away from the wall and face nailed near the wall with 8d casing nails or cut flooring nails (see Fig. 11-5). After the first row is in place the second row of flooring is pulled tightly against it and blind nailed. Care must be taken not to damage the top edge or the tongue of the floor boards when edge nailing. The nails should be placed at the tongue and should make an angle of about 45° with the subfloor. To set the nails and also drive the boards tightly together a nailset is placed on the tongue, as shown in Fig. 11-6, and hammered.

After a few rows of boards are in place many floor layers use an automatic strip-floor nailing machine (see Fig. 11-7). This nailer uses nail clips of special flooring nails. The machine is placed over the tongue of the board and with one blow of the special hammer the board is drawn up tight and the nail set below the surface of the tongue. Some floor-nailing machines are operated with compressed air.

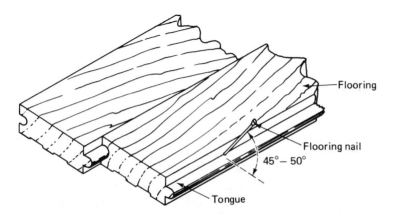

Location and angle of nails

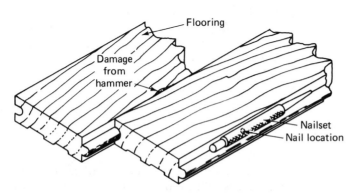

Setting nails to avoid hammer damage

Fig. 11-6   Nailing Strip Flooring

Fig. 11-7   Floor Nailing Machine (Courtesy Rockwell Manufacturing Company)

Sanding and Finishing   Following the installation of wood floors the interior trim is applied. The sanding and finishing of the floors is not accomplished until nearly all other finishing work in the building is completed, to avoid having the floor finish damaged during the final work of completing the house.

The floors are sanded by specialists using large drum floor sanders for the main part of the floor and small rotary edgers for spaces near the walls where the large sanders cannot be used. Inside corners and areas near door jambs and trim which the edger cannot reach are finished with a scraper.

When the sanding is completed the floors are swept clean and the desired floor finishes are applied. Two or more coats of finish are needed, and the floor should be barricaded to avoid damage while the finish is drying.

PARQUET FLOORING

Parquet block flooring is made from a number of hardwoods and in a variety of patterns. The most commonly used block floor is made of four strips of tongue-and-groove strip flooring

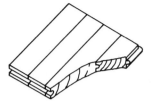

Unit parquet block

Laminated block (plywood)

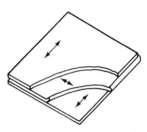

Strips bonded to membrane

Fig. 11–8   Types of Parquet Flooring

held together by metal splines on the back. These blocks are usually 9″ square.

A second type of parquet block flooring is made of square-edged strips bonded to a membrane. The membrane is usually a heavy brown paper and the strips are usually bonded in a criss-cross pattern. The membrane is removed after the flooring is installed.

The third type of parquet block flooring is the laminated block or plywood parquet block. It is made of three plys, with the grain of the center ply running at right angles to the grain of the face plys. The ends and edges are tongue and groove. The three main types of parquet flooring are illustrated in Fig. 11–8.

Installing Parquet Block Flooring

The surface to which parquet block flooring is applied must be reasonably smooth and free from irregularities. It cannot be installed over a board subfloor, but it can be satisfactorily placed on plywood subfloors and on dry concrete floors.

Before attempting to install parquet flooring the subfloor should be swept clean and made free of dust and irregularities. Following the manufacturer's recommendations a sealer is applied over the clean subfloor. Not all adhesives require a sealer, and some require sealers to obtain a good bond only under certain conditions. This is why it is important to follow the adhesive manufacturer's recommendations on the use of sealers.

It is necessary to make a layout to determine the starting points when installing parquet blocks, in order to maintain an even border on the perimeter at opposite sides of the room. The layout should be made in such a manner that the width of blocks along the wall will not be less than one-half the width of a full block.

To determine the starting lines the room is measured, and the size is converted from feet and inches to total inches. Next the dimensions are divided by 2 to locate the center lines. After the center lines are established the distance from the wall to the center line is divided by the width of a parquet block to determine the width of the end blocks when starting at the center line. If the width of the end row is less than one-half the width of the block, the center line is made to pass through the center of the block (see Fig. 11–9).

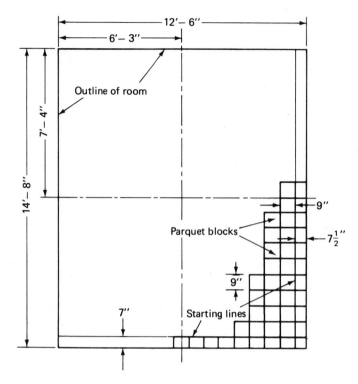

Fig. 11–9 Establishing Starting Lines for Parquet Flooring

EXAMPLE

Room size: 12'6" by 14'8"

Size of parquet block: 9" by 9"

12'6" = 150"

150" ÷ 2 = 75" from wall to center line

75" ÷ 9" = 8⅓ blocks from center line to wall

Note: This is unacceptable because the end block is less than one-half the width of the block. Therefore, the center line must pass through the center of the block.

75" – 4½" (one-half the block width) = 70½"

70½" ÷ 9" = 7 full blocks plus 1 block 7½" wide on each side of the center row of parquet blocks

14'8" = 176"

$$176'' \div 2 = 88''$$

$$88'' \div 9'' = 9 \text{ full blocks plus 1 block } 7'' \text{ wide}$$

After the widths of the end rows are determined starting lines should be snapped on the floor. These chalk lines are followed when putting the first row of blocks in place. Care must be taken not to cover these starting lines completely when applying the mastic.

Applying Sealer and Mastic   Sealers, when used, should be compatible with the mastic that will be used to secure the floor. The purpose of the sealer is to bond small particles of the subfloor to itself and to increase the bond of the mastic to the subfloor. The sealer also makes the mastic easier to apply. All sealers should be applied according to the manufacturer's instructions.

Mastics should be applied over a subfloor which has been swept clean or one which has been properly prepared with a sealer. In all cases the mastic should be applied according to the directions supplied by the manufacturer. Sealers should be completely dry before the mastic is applied.

The mastic is spread with a notched trowel (see Fig. 11–10) to the recommended coverage. The size and spacing of the notches in the trowel regulate the amount of mastic applied to the subfloor. Care should be taken to use a trowel with proper-size notches. If the notches are too small there will not be enough mastic applied to the floor, the result of which will be parquet blocks which loosen and become unfastened.

When the notches in the trowel are too big an excessive amount of mastic will be applied. Large amounts of mastic require a longer drying time and also have a tendency to seep upward between the joints of the parquet blocks. Mastic between the edges of individual blocks detracts from the appearance of the floor and can cause a maintenance problem, because the mastic will continue to seep until the excess is worked out.

Placing Parquet Blocks   Parquet blocks are placed in accordance with previously established layout lines. These chalk lines are snapped on the floor before the mastic is applied. Care is taken not to obliterate these lines completely when the mastic is applied. Individual blocks are placed along the layout line and turned alternately to establish the desired pattern. The individual blocks are placed close together but are not forced tight. The small space allowed between the individual blocks provides

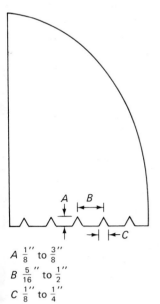

$A \; \frac{1}{8}''$ to $\frac{3}{8}''$

$B \; \frac{5}{16}''$ to $\frac{1}{2}''$

$C \; \frac{1}{8}''$ to $\frac{1}{4}''$

Fig. 11–10   Notched Trowel Design

room for expansion of individual blocks due to changes in moisture content. If the blocks are laid too close together they will expand and break loose from the mastic, with a bulge or hump resulting in the floor.

Parquet blocks around the perimeter of the room are marked for cutting and are cut to fit as required. If prefinished blocks were installed the floor is ready to use after the scraps and sawdust have been swept up and removed. When unfinished material was used it is necessary to sand and finish the floor in the same manner as described for strip flooring.

UNDERLAYMENT FOR RESILIENT FLOORS

Underlayment provides a smooth, stable, durable base for resilient floor coverings and wall-to-wall carpeting. The material used for underlayment should be of uniform thickness, since variations in thickness will cause uneven joints that will show through the floor covering.

It is important that the underlayment be properly fastened to the subfloor and the joists to prevent movement in the underlayment that would be reflected in the finish floor covering. Proper fastening of the underlayment also provides for a stiffer floor which will resist the squeaking that results from loose nailing and bending in the floor system.

Most underlayment is made from one of three types of materials. These are untempered hardboard, particleboard, and plywood. All are available in 4' or 8' sheets.

Hardboard is usually used in ⅛" thickness for remodeling work. It is applied over solid board floors to provide a smooth base for resilient flooring materials.

Particleboard is manufactured in ⅜", ½", ⅝", and ¾" thicknesses. When applied over a properly prepared subfloor it gives a smooth, uniform surface with no voids or grain blemishes. It is hard and durable and resists penetration by furniture legs or spike heels.

Plywood is manufactured in a special underlayment grade with a plugged "C"-grade veneer on the face which has been fully sanded. The veneer immediately below the face surface is also of "C" grade to provide the necessary strength to resist the concentrated loads imposed by furniture, spike heels, and so on. The remaining veneers are "D" grade for economy.

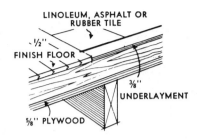

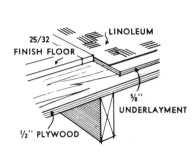

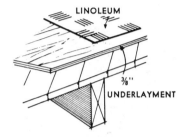

Fig. 11-11 Underlayment Installation (Courtesy American Plywood Association)

Underlayment-grade plywood is manufactured in ¼", ⅜", ½", ⅝", and ¾" thicknesses.

The thickness of underlayment to be used will be determined by various job conditions. On new residential work where the underlayment is applied over a subfloor the thickness is regulated by the thickness of other flooring materials. If finish flooring ²⁵⁄₃₂" thick is used the underlayment would be ⅝". With a ½" finish floor the underlayment would be ⅜" thick. The ⅛" difference in thickness between the finish flooring material and the underlayment allows for the thickness of resilient floor covering. Where there is no other finish flooring material with which to match up, the ⅜" material can be used throughout the building (see Fig. 11-11).

Installing Underlayment

Subflooring should be swept clean and all humps and uneven areas should be leveled. Any necessary repairs should also be made, and the entire subfloor should be checked to see that it is nailed adequately. Any nails which are pulling loose should be driven home, and extra nails should be installed to be sure that the subfloor will not loosen.

The location of all the joists should be marked on the wall at the base line. If building paper or deadening felt is used it is

Fig. 11-12   Installing Underlayment (Courtesy American Plywood Association)

applied at this time. Sheets of underlayment are installed so that all end joints fall at the center of a joist, and the end joints in adjacent rows of underlayment are staggered.

The usual procedure in installing underlayment is to start with a full sheet by aligning its edge parallel to the wall and the ends at the center of the joists as indicated by the marks on the wall. When the sheet is aligned the ends are nailed in place, and a few nails may be placed in the field of the panel to draw it tight to the floor. Other panels are aligned with the first panel and tacked in place (see Fig. 11-12). Pieces are cut to fit into the corners of the room and elsewhere as required. After the floor in the entire room is covered with underlayment, it is nailed completely in place according to standard specifications.

Nailing and Stapling   For particleboard ⅜″ through ⅝″ thick 6d nails are recommended, and 8d nails should be used on ¾″ particleboard. Nails should be spaced a maximum of 6″ on center along the edge of the panels and 10″ on center in the field of the board. Nails along the ends of the board should be placed ½″ to ¾″ away from the edge.

If staples are used on particleboard they should be 1⅛″ long for ⅜″ underlayment, 1½″ long for ⅝″ underlayment, and 1⅝″ long for ¾″ underlayment. Staples should be spaced 3″ apart along the edges and 6″ apart in the field of the panel.

Underlayment nails with at least 20 annular rings per inch and a 5⁄16″ head should be used on plywood underlayment. These nails should be long enough to penetrate at least ⅝″ into or completely through the subfloor. Longer nails which penetrate the joists are desirable, and 6d and 8d nails are often used. Sixteen-gauge staples of sufficient length may also be used.

Nails should be spaced every 6″ along plywood panel edges and every 8″ in the field of the panel, and staples should be spaced every 3″ along the edges and every 6″ in the field of the panel.

After the nailing is complete the scraps and sawdust should be swept up and removed from the building. Any leftover material should be put in an out-of-the way place or removed to make room for other work.

REVIEW QUESTIONS

1. What are the desirable characteristics of wood floors?
2. What are the three main categories of flooring installed by carpenters?
3. Outline the procedure for installing strip flooring.
4. What is the purpose of felt papers between subfloors and finish floors?
5. How is end-matched flooring fastened in place?
6. When are finish floors sanded?
7. How is sanding done near walls and in corners?
8. Outline the procedure for installing parquet flooring.
9. When are sealers needed when installing parquet flooring?
10. How is mastic applied?
11. Why should parquet blocks not be installed too closely together?
12. What types of underlayment are in common use? Describe each.
13. Outline the procedure for installing underlayment.
14. What are some of the nailing requirements for underlayments?

interior trim

12

Interior trim is often called millwork. Sometimes it is called finish carpentry. It involves the installation of door jambs, moldings, cabinets, doors, and hardware necessary to finish the building.

Many different species of lumber are used to manufacture interior trim. The most commonly used lumber includes the various western pines, redwood, oak, and birch. Interior trim is also made from walnut, cherry, maple, mahogany, and other species of lumber on special order.

BUILDING PREPARATION

Before any millwork is delivered to a job for installation the building should be thoroughly dried out. This is particularly important if the walls are plaster or if there is a concrete floor. If sufficient heat and ventilation have not been provided to dry these materials before the millwork is delivered the excess moisture will be absorbed by the millwork. If the millwork is fitted and installed while in a moisture-laden condition it will dry out and shrink as the building dries out. This shrinkage results in poor-fitting joints in the various trim members. To avoid the opening of fitted joints, trim should be delivered only after the building has dried out. It should also be allowed to lie in the building for a few days before installation to attain equilibrium with the air moisture content present in the building.

DOOR JAMBS

Door jambs are interior door frames. They are made of finish lumber and are installed in the rough opening to provide a straight and true opening of proper size to receive the door. Most jambs are made from lumber ¾″ thick. The width of the jamb depends on the total thickness of the finished wall, and a number of "standard" widths are available. Most door jambs (not all) have the edges slightly tapered so that the face side of the jamb is wider than the wall (see Fig. 12–1). This slight taper

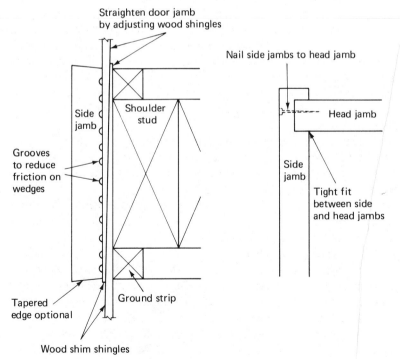

Fig. 12–1   Door Jamb

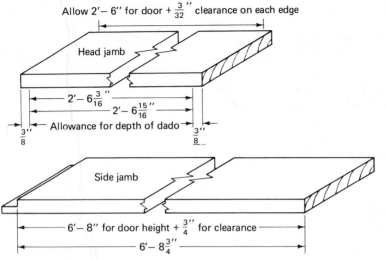

Fig. 12-2 Sizing Jamb Parts

Level held in level position in front of door opening

Door opening

Floor line

Amount to add to jamb length on this side of door to maintain level head jamb

Fig. 12-3 Checking Floor Level

aids in obtaining a tight fit between the door jamb and the door casing.

Before door jambs can be installed they must be cut to size and assembled. The head jambs must be cut long enough to allow for the depth of the dado in the side jambs plus a small allowance for door clearance. The depth of the dado is usually $\frac{3}{8}''$, but it should always be checked to be sure. The head jamb for a $2'6''$ door would be cut $2'6\frac{15}{16}''$ long. This allows $\frac{3}{8}''$ on each end for the dadoes, $\frac{3}{32}''$ on each side for clearance between the jamb and door, and $2'6''$ for the door itself (see Fig. 12-2).

The side jambs must be cut long enough to allow for door height plus $\frac{1}{2}''$ to $1''$ clearance between the floor and the bottom of the door. This clearance is needed to allow for throw rugs and carpeting.

The length of the side jambs in Fig. 12-2 is $6'8\frac{3}{4}''$ measured from the bottom of the jamb to the dado. This allows $6'8''$ for the door and $\frac{3}{4}''$ for clearance.

If the floor is out of level in the door area it is necessary to cut one side jamb longer than the other. The additional length is determined by placing a level on the floor across the door opening and measuring the distance the level must be raised at one end to reach a level position (see Fig. 12-3).

The side jambs and head jamb are placed on a flat working surface (usually the floor) with the head jamb aligned in the

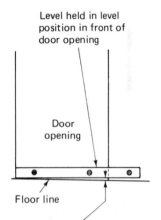

dado. To assemble the jambs 6d box nails are driven through the side jambs into the head jamb (see Fig. 12-4). Care should be taken to get a good fit between the head jamb and the dado at the face of the side jamb. Open joints at this point give the appearance of poor workmanship.

Jamb Installation

The assembled jamb is placed in the rough opening and held in place temporarily, either by nails or wedges placed near the head jamb between the rough framework and the side jamb. Then the head jamb is checked for level. If it is not level the length of the side jambs is adjusted as required.

To keep the side jambs properly spaced at the bottom of the opening a spreader is placed on the floor between the jambs.

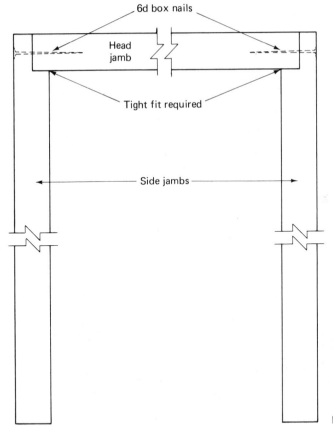

6d box nails

Head jamb

Tight fit required

Side jambs

Fig. 12-4   Assembled Jamb

The length of this spreader is exactly the same as the distance between the side jambs as measured at the head jamb.

One side jamb is plumbed with a level and straight edge. Shim shingles are adjusted at the top and bottom ends to bring the jamb to a plumb position. When the jamb has been plumbed 8d casing nails or 8d finish nails are driven through the jamb and shingles to anchor the jamb to the framework.

The jamb is straightened between the top and bottom ends with the aid of the straight edge. The straight side of the straight edge is held against the jamb, and shim shingles between the jamb and rough framework are adjusted to straighten the

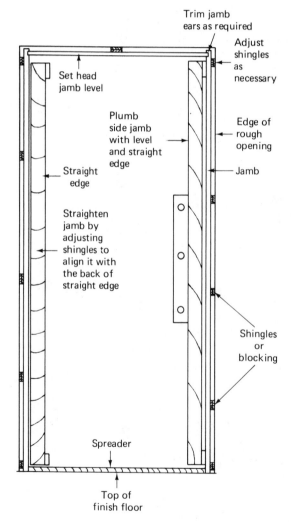

Fig. 12-5   Jamb Installation

jamb (see Fig. 12-5). After the jamb is straightened, nails are driven through the jamb and shingles. When nailing is complete the jamb is rechecked with the straight edge, and any necessary adjustments are made.

The second side jamb is plumbed by bringing the lower end against the spreader and installing the necessary shims. The jamb is held in place by nailing through the shims into the framework. The second side is straightened in the same manner as the first side jamb.

Necessary shims are also installed and nailed in place to keep the head jamb in alignment. When placing shims around the perimeter of the jamb care should be taken to place shims behind the hinge areas and the lock strike plate area. Shims at these points provide solid backing for the door hardware. All shims are cut back to the wall line after the jamb installation is completed.

DOOR CASINGS AND DOOR STOPS

The purpose of the door casing is to cover the joint between the door jamb and the wall. Casings are usually made of the same material as the jamb and are available in a variety of standard patterns (see Fig. 12-6).

The door stop, as its name indicates, serves to stop the door and prevent it from swinging through the opening. Stops are made from the same kind of lumber as the jambs. Door stops vary in size and shape, but most stops are ½″ thick.

Installing Door Stops

Most carpenters prefer to cut door stops to length before installing door casings because it is easier to mark the length on the stops before the casings are in place. The ends of molded door stops are mitered where the side and head stops meet. The head stop is usually cut to length first and the side stops cut later. The cut pieces are usually nailed to the jambs temporarily to keep them from becoming lost during the finishing operations.

The stops are nailed permanently in place after the doors have been hung. The head stop is located so that the door stops

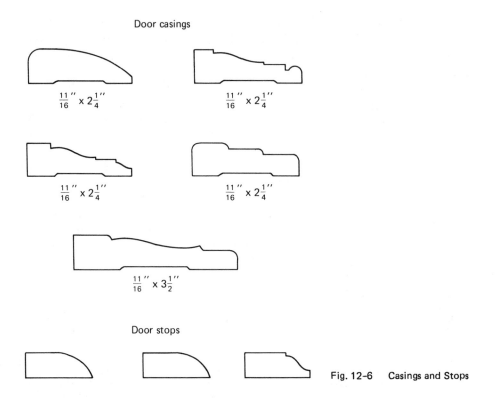

Door casings

$\frac{11}{16}'' \times 2\frac{1}{4}''$

$\frac{11}{16}'' \times 2\frac{1}{4}''$

$\frac{11}{16}'' \times 2\frac{1}{4}''$

$\frac{11}{16}'' \times 2\frac{1}{4}''$

$\frac{11}{16}'' \times 3\frac{1}{2}''$

Door stops

Fig. 12-6   Casings and Stops

at the proper point. The stop along the hinge side is nailed in place so that it will not bind on the edge of the door when it is being closed. Sufficient clearance is allowed for wood finishes which will be applied to the door and stops.

The stop on the lock side of the door is often installed after the lock and strike plate are in place. This allows the carpenter to nail the stop in position along the door while providing sufficient clearance to afford ease of operation.

Installing Casings

Casings are installed with a ⅛" to ¼" margin along the edge of the door jamb. Molded casings are fitted with a mitered joint where the head and side casings meet. Casings of rectangular pattern are fitted with a butt joint (see Fig. 12-7). The general procedure of cutting, fitting, and installing casings is discussed in the following paragraphs.

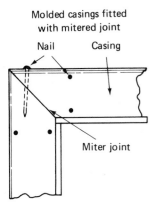

Molded casings fitted
with mitered joint

Nail   Casing

Miter joint

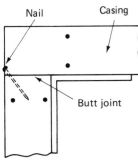

Nail   Casing

Butt joint

Rectangular casings fitted
with butt joint

**Fig. 12-7   Fitting Side and
Head Casings**

The lower end of the casing should be cut square so that it will fit properly along the floor. To obtain the proper length for the side casing it is held in place along the jamb and marked for length at the head jamb. Miter cuts are made in accordance with the marks. When mitered to length one side casing can be nailed in place. It is common practice to use 8d finish nails on the thick edge of the casing when it is fastened over plaster or drywall. The thin edge is fastened to the jamb with either 3d or 4d finish nails. All nails are set below the surface of the casing.

One end of the head is usually fitted before the second side casing is installed. This method allows the carpenter the opportunity to trim the miter cut on the head casing to get a better fit without fear of making the head casing too short. After the miter fits to satisfaction the casing is marked for length, and the second miter cut is made.

The second side casing can now be installed in a manner similar to that for the first casing. The head casing is fitted between the side casings and nailed in place.

WINDOW TRIM

A typical window will have a number of different kinds of trims. These trim items will include a stool, apron, stops, casings, and mullion casings (see Fig. 12-8). Some types of window frames will also require a subjamb or jamb extension.

Sometimes the window is trimmed to look like a picture frame. When this is done the stool and apron are omitted, and the casings and stops are placed around all four sides (see Fig. 12-9). If the window has a sloping sill at the point where the stool is normally applied a wedge-shaped filler strip must be installed to provide a flat surface for the attachment of the window stop.

Installing Window Trim

The procedure followed when installing window trim will vary with the type of window frame and the preference of the installer. In general, the procedure for trimming a window with a stool will be as indicated in the following paragraphs.

First the length of the stool must be determined by mark-

Fig. 12-8   Window Trim – Stool and Apron

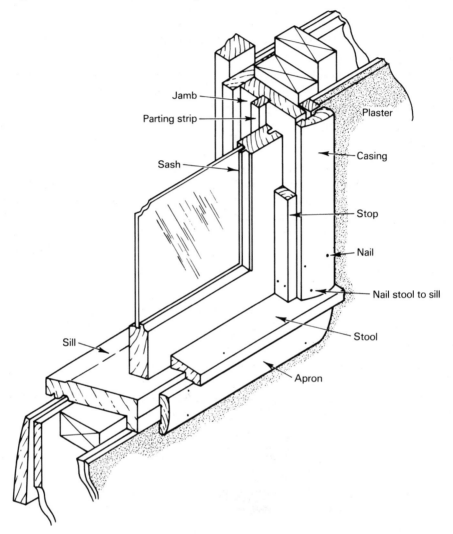

ing the width of the casing and return on each side of the window frame (see Fig. 12-10). The amount of return is usually equal to the thickness of the casing, but it can be varied to meet the requirements of the job.

After the stool length is marked on the wall a piece of stool is held in place along the window sill, and the length marks are transferred to it. The ends of the stool are also scribed to the wall, and the stool is cut along the layout lines.

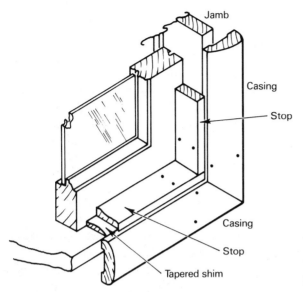

Fig. 12-9   Window Trim – No Stool

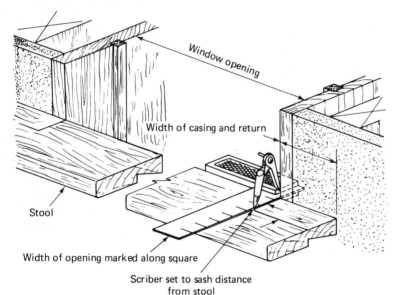

Fig. 12-10   Window Stool Layout

Ends of the stool may be cut square, or they may be coped to the profile of the molded edge of the stool.

The stool is fastened to the sill with 8d finish nails and is set square to the window sash by checking with a combination square. An apron is cut to length and fastened below the stool. Its length runs from the far sides of the side casings (see Fig.

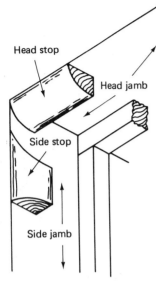

Head stop

Head jamb

Side stop

Side jamb

Fig. 12-11   Coped Stops

12-10). The function of the apron is to cover the joint between the sill of the window frame and the interior wall material.

Side casing material is squared on the lower end and held in place along the window frame for length marking. After marking, the side casings are mitered, and one end of the top casing is mitered in preparation for fitting it to the side casing.

Before the side casings are installed most carpenters will mark the length on the window stops, because it is easier to do than after the casings are in place. This is accomplished by setting the squared end of the stop on the stool in a position parallel to the side of the frame and marking the length at the head jamb. The head stop is held in place along the head jamb and marked for length.

It is common practice to cut the ends of the head stop square and to cope the upper ends of the side stops to fit around the head stop (see Fig. 12-11).

Different procedures can be followed when installing window casings. However, many tradesmen prefer to install one side casing and then fit the end of the head casing to it before marking the head casing to length. Following this procedure can save a lot of recutting if the frame is not square. After the joint fits properly the casing is marked for length and mitered according to the marking.

The second side casing is usually installed at this time, and the head casing is fitted between the two side casings. If any fitting is required at the miter of the second casing and head casing it is accomplished by trimming with a block plane or molding shear in the same manner as the first mitered end.

When all casings have been nailed securely in place, the head stop is installed, and the coped side stops are fitted to it and nailed in place. Care should be taken to avoid placing the stops too close to the sash which can be opened, because this would cause the operating sash to bind. All nails are set below the surface of the trim members.

BASE TRIM

Base trim is generally made from the same lumber as other trim used in the room. One exception is that when hardwood trim is used in rooms with closets, the closet trim is usually a softwood.

There are many patterns of baseboard available (see Fig.

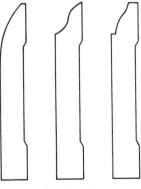

Base boards

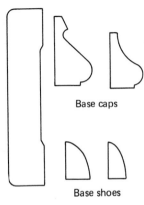

Base caps

Base shoes

Fig. 12-12   Base Materials

12-12). Some of these base materials may be combined, while others are used separately. When base materials are used separately they form a single-member base. The single-member base (Fig. 12-13) is often used when carpeting will be used in the room. If it is installed over a hardwood floor it must be fitted to the irregularities of the floor. Any space between this type of base and the floor is poor in appearance and becomes a dirt collector.

The two-member base (Fig. 12-14) is used in most installations. It can be made from a number of different base patterns combined with a shoe molding. The purpose of the shoe molding is to cover the joint between the baseboard and the floor. A variation of the two-member base used with carpeting is made by removing the shoe molding from the three-member base.

The three-member base (Fig. 12-15) is made up with a baseboard, base cap, and base shoe. It is usually used on plastered walls, but it can be used with any wall material. The baseboard may vary in width from 4″ to 8″ or more, and the narrow base cap installed over the base may be made to follow irregularities in the wall that the baseboard would bridge. As with the two-member base, the shoe molding covers the joint between the floor and the baseboard.

Fig. 12-13   Single Member Base

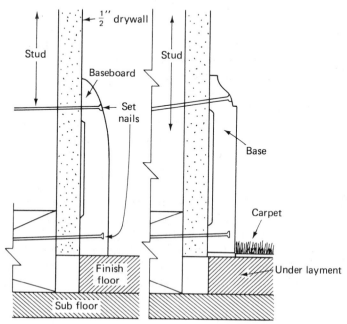

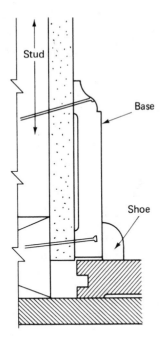

Stud

Base

Shoe

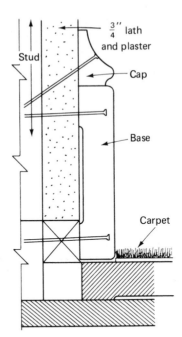

$\frac{3}{4}''$ lath and plaster

Stud

Cap

Base

Carpet

Fig. 12-14   Two Member Base

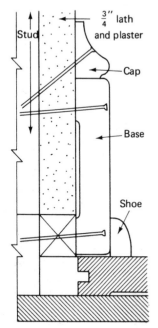

$\frac{3}{4}''$ lath and plaster

Stud

Cap

Base

Shoe

Fig. 12-15   Three Member Base

Installing Base Trim

Before the base trim can be installed the floor near the wall must be swept clean, and the location of the studs must be marked on the floor. The studs are usually located by the nails in the ground strips or the nails at the lower edge of the wallboard.

The first piece of baseboard to be installed is cut to run from corner to corner of the room or from the corner to a door casing. Care must be taken to fit the baseboard to the door casing. In most cases a simple square cut will suffice, but sometimes it is necessary to trim the end of the base or cut it at an angle to obtain a proper fit with the casing.

Either 6d or 8d finish nails are used to fasten baseboards in place. The length of the nail used will depend on the thickness of the baseboard and other materials which must be penetrated before the nail reaches the stud or plate. All nails should be driven into the studs as indicated by the marks placed on the floor before the baseboard was put in place. These nails are set below the surface of the trim members.

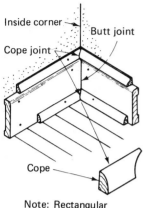

Note: Rectangular
patterns may
be butted,
molded patterns
must be coped

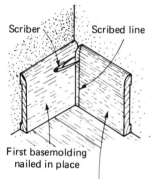

Fig. 12-16   Coped Base –
Inside Corner

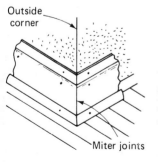

**Fig. 12-17   Mitered Base
Outside Corner**

Fitting Inside Corners   Where walls come together to form an inside corner molded baseboards are coped (see Fig. 12-16). This type of joint gives the appearance of a miter but is easier to fit and does not open up to show a poor joint as readily as a mitered joint would if used at this point.

To make a coped joint the base to be coped may be scribed by setting it near the base already installed (see Fig. 12-16). A coping saw is used to cut along the scribed line. In following the scribed line, back wood is removed to aid in providing a tight joint and a good finished appearance.

An alternative to scribing the baseboard is to cut the end at a 45° angle as for a mitered joint and then cope along the profile of the molding as outlined by the miter cut.

Fitting Outside Corners   Outside corners are always mitered (see Fig. 12-17). The base is fitted at one end, and the length is marked at the outside corner. After mitering the base is fastened in place, and the meeting base with a mitered end is fitted to it. After fitting, this second piece of base is marked for length and cut as required.

CLOSET SHELVING

Shelving in clothes closets is usually made from a 1 by 12 board. This board is usually supported on a hook strip which is installed around the perimeter of the closet. If the shelf is over 4′ in length a central shelf and pole support is also installed. Expandable metal shelving with all necessary attaching hardware is also used in closets.

Installing Shelves and Closet Poles

Shelves in clothes closets are usually installed 5′6″ above the floor (see Fig. 12-18). The hook strip which supports the shelf is leveled and nailed to the studs with 6d or 8d finish nails. Before the shelf is cut to length the ends of the closet should be checked for squareness. The ends of the shelf should be cut to the shape of the end walls. Care should be taken not to cut the shelf too long, because a long shelf would wedge in place, damage the walls, and be difficult to remove. A shelf which is cut

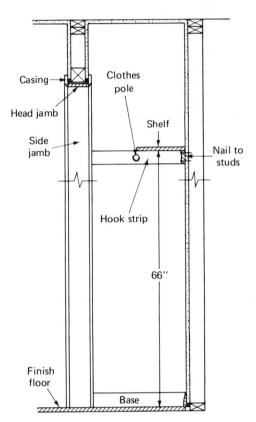

Casing

Clothes pole

Head jamb

Side jamb

Shelf

Nail to studs

Hook strip

66″

Finish floor

Base

Fig. 12-18   Closet Shelf and Pole

too short would not have sufficient support at the ends and would be in danger of slipping off the supporting hook strip.

When required a shelf and pole support bracket is used. This bracket is attached to the wall studs with long wood screws. Its function is to support long closet shelves and poles to prevent undue sagging.

Closet poles are supported by rosettes which are usually fastened to the hook strip. These rosettes may be made of wood, metal, or plastic. Depending on the type, wood rosettes may be fastened by nailing or by a single wood screw. Metal rosettes are usually fastened in place with one to three wood screws. Plastic rosettes are fastened with a single wood screw.

The closet pole is usually made of wood 1¼″ in diameter. This pole is carefully cut to length and put in place between the rosettes. When 1″ steel pipe is used for the clothes pole, extra care in measuring and cutting is necessary as it is difficult to remove a small amount if the pipe is cut too long.

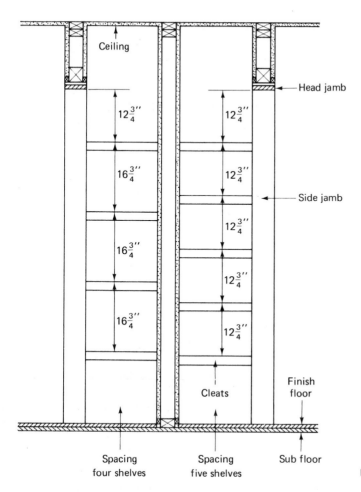

Fig. 12-19   Linen Closet Shelving

When expandable metal closet poles are used no cutting to length is necessary. However, care must be taken to install all attaching hardware in accordance with the instructions provided. Failure to install necessary locking screws can result in the pole coming down during use.

Installing Shelves in Linen Closets

Shelves in linen closets may be made from boards or plywood, or they may be of the expandable metal type. Most linen closets will be arranged for either four or five shelves (see Fig. 12-19).

When four shelves are used the shelf supports are spaced so that there will be 16″ between shelves. It is common practice to locate the top shelf supports 12¾″ down from the head jamb of the closet door.

If five shelves are used the shelf supports are spaced so that there will be 12″ between shelves. As with a closet having four shelves, the top shelf support is located 12¾″ down from the head jamb of the closet door (see Fig. 12-19).

Wood shelves are cut to size, and the location is marked on the back edge. Marking the location is necessary because often the walls are not parallel, and it becomes necessary to cut each shelf a different size.

CABINETS

Kitchen cabinets are almost always prefabricated in a mill or cabinet shop. They may be made up from the standard stock units, or they may be custom built. In either case they are delivered to the job in preassembled sections, and the carpenter will install them in accordance with the plan layout.

Typical kitchen cabinets are made up with a base cabinet and a wall cabinet (see Fig. 12-20). The base cabinet is usually 36″ high and 24″ deep. The counter top is usually made with a plastic laminate bonded to plywood or particleboard. Counter tops usually project 1″ beyond the face of the cabinet, and the back splash may project 4″ or more inches above the counter.

Wall cabinets vary in height and depth. Most wall cabinets are 30″ to 32″ in height and contain two movable shelves. The depth varies greatly and can be made to accommodate various requirements. An inside depth of 12″ or more is desirable.

It is not the carpenter's job to design cabinets. However, on occasion he is asked for advice on how the cabinets should be built. The most important points to consider are what the cabinet is to be used for, and who will use it. With this in mind it should be fairly easy to answer questions as to size and depth of the cabinets.

An important point to consider is the height of the counter work area. The "standard" height is 36″, but this is too high for women of average height, and a counter height of 30″ to 32″ would be more desirable as illustrated in Fig. 12-21.

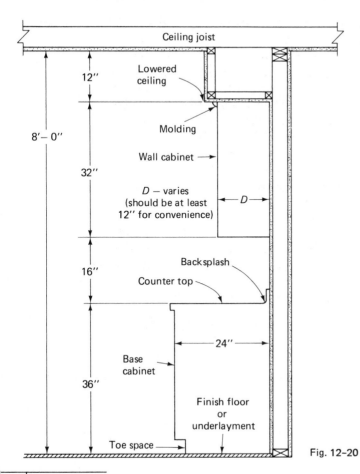

Fig. 12–20   Kitchen Cabinet – Typical Section

Optional cabinet
seasonal storage

13″

Average height 5′– 3″
Spice
cabinet

15″ to 18″

24″   33″

Build the kitchen to
fit the housewife

**12-21   Kitchen Cabinet
Design Considerations**

Installing Kitchen Cabinets

Kitchen cabinets should be installed tight to the wall and fastened securely in place. Some carpenters prefer to install the base cabinets first, while others prefer to install the wall cabinet first. The advantage of installing the wall cabinet first is that it is easier to reach than after the base cabinet is in place. However, the disadvantage is that the cabinet must be raised from the floor to its wall location, but when the base cabinet is in place the wall cabinet is placed on the base cabinet first and then raised to proper position.

Before fastening, wall cabinets are held in position on temporary supports. These supports may be simple T posts, or they may be "saw horses" built to the proper height. With the cabinet held in place on the supports it is scribed to the wall as required. If any trimming is required the cabinet is removed from the temporary support and fitted as necessary.

When properly fitted the wall cabinet is placed back on the temporary supports and fastened into place with nails or screws installed through the fastening strip provided at the back of the cabinet (see Fig. 12-22). It is important that the cabinet be fastened directly into the studs to obtain adequate support. Nails or screws which penetrate only drywall or plaster will not adequately support the cabinet when loaded with dishes, appliances, food, and other items.

Base cabinets should be installed level in both directions, and they should also be fitted to the wall. The cabinet is adjusted for level with wood shim shingles and checked in both directions. When properly located and leveled it is fastened to the wall by nailing through the top rail. The cabinet is also fastened to the floor by toenailing through the side of the cabinet with finish nails (see Fig. 12-23).

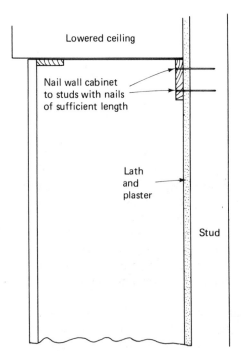

Lowered ceiling

Nail wall cabinet
to studs with nails
of sufficient length

Lath
and
plaster

Stud

Fig. 12-22   Fastening Wall Cabinet

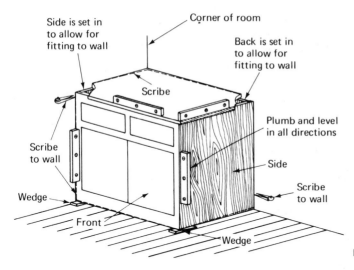

Fig. 12-23   Installing Base Cabinets

DOORS

Doors manufactured from various kinds of lumber may be obtained. Most of these used in residential construction are of flush design; that is, they are flat surfaced as compared to louvered or paneled doors.

Flush doors are manufactured of either hollow-core or solid-core construction. Hollow-core doors are manufactured in 1⅜″ and 1¾″ thicknesses. The edges and ends of the door are made of solid wood approximately 2″ wide. This frame is covered on each side with ⅛″-thick plywood. To support the plywood in the area between the edge framework various types of spacers are used. Many of these spacers resemble egg crating and are often referred to as such.

Solid-core doors used in residential construction are made with a staved-lumber core. This staved core gives the door its name (see Fig. 12-24). Other types of solid-core doors are available, but they are seldom used in residential construction.

Determining the Hand of a Door

The hand of the door is determined by facing the door from the outside of the room or building. If the hinges are on

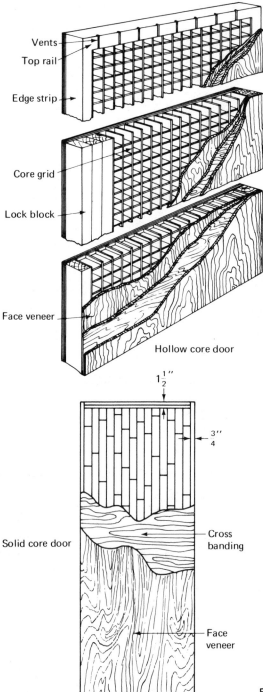

Vents

Top rail

Edge strip

Core grid

Lock block

Face veneer

Hollow core door

$1\frac{1}{2}''$

$\frac{3}{4}''$

Solid core door

Cross banding

Face veneer

Fig. 12-24   Door Construction

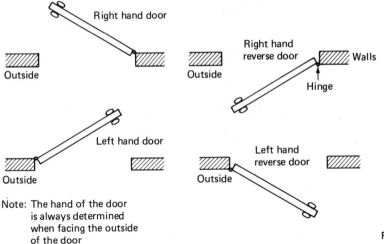

Note: The hand of the door
is always determined
when facing the outside
of the door

Fig. 12-25   Hand of Doors

the right-hand side and the door swings in it is a right-hand door, but if the door swings out it is a right-hand reverse door. That is, it has a reverse bevel.

If the hinges are on the left-hand side the door is left hand if it swings in, but when it swings out it is a left-hand reverse door (see Fig. 12-25).

It can be argued that there are only two hands of doors, and for some types of hardware installations the argument is successful. However, to be sure that the hardware supplier will supply the proper type of locks and other hardware it is best to identify doors in the manner outlined in the previous paragraphs.

Exterior Doors

Exterior doors are fitted to the head and side jambs of the door frame before the threshold is installed. As a general rule these doors are fitted with a $\frac{1}{16}''$ clearance on the top edge and hinge side of the door. A clearance of $\frac{3}{32}''$ is allowed on the lock side. On doors $1\frac{3}{4}''$ thick a bevel of $\frac{1}{16}''$ on the lock side will provide clearance between the jamb and door (see Fig. 12-26).

The threshold is installed after the lower end of the door is marked for cutting. The usual procedure for marking the lower end of the door is to set a scriber to the thickness of the threshold and to scribe the door when it is closed. When the door is in a closed position over the installed threshold, the out-

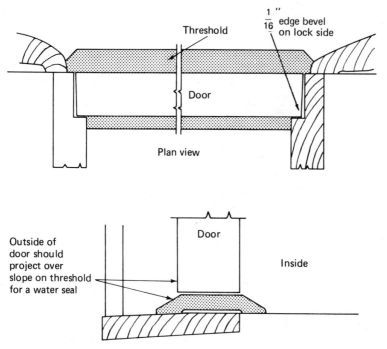

Fig. 12-26   Exterior Doors

side face of the door is over the beveled portion of the threshold (see Fig. 12-26). With the threshold installed in this manner it provides an effective stop against rain water.

Interior Doors

Interior doors are fitted to the head and side jambs in the same manner as exterior doors. Most interior doors are of hollow-core construction and $1\frac{3}{8}''$ thick. A $\frac{1}{16}''$ clearance at the top and the hinge side is usually sufficient. A clearance of $\frac{3}{32}''$ between the jamb and door is allowed on the lock edge of the door. No bevel on the lock side is necessary for doors $1\frac{3}{8}''$ thick.

A $\frac{1}{2}''$ to $1''$ clearance is left between the bottom of the door and the floor. The amount of clearance is usually governed by the thickness of carpeting and throw rugs which will be placed on the floor. Thresholds are seldom used with interior doors.

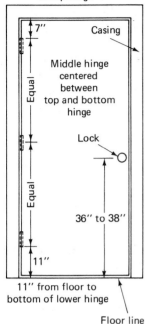

7" from top of door
to top hinge

7"

Casing

Middle hinge
centered
between
top and bottom
hinge

Equal

Lock

Equal

36" to 38"

11"

11" from floor to
bottom of lower hinge

Floor line

**Fig. 12-27   Locating Door
Hardware**

Installing Doors

After fitting the door the hinges are located. The top of the top hinge is usually located 6″ to 7″ from the top of the door. The bottom of the lower hinge is usually located 10″ to 11″ above the floor. This distance will vary with local practice. When a center hinge is used on exterior doors, it is centered between the top and bottom hinges (see Fig. 12-27).

Butt hinges may be installed with the aid of a hinge template and router, or they may be marked with a butt gauge and gained or mortised with a chisel. When a hinge template is used the manufacturer's instructions should be followed carefully.

When the gaining is done by hand a series of chisel cuts about $\frac{1}{16}$″ to $\frac{1}{8}$″ apart should be made to a depth equal to hinge thickness (see Fig. 12-28). These small chips can be pared away easily and any irregularities can be cut away to provide a flat surface for the hinge.

Various types of door hardware are available for pocket doors, bypassing doors, and bifolding doors. The carpenter should read and follow the instructions provided by the hardware manufacturer since each brand of hardware has different features. Following the instructions provided makes installation easy and avoids costly mistakes.

HARDWARE

The carpenter is called upon to install many different hardware items. Among these are hinges, pulls, and catches on cabinets. Window sashes often require the application of locks and lifts, doors require locks and door bumpers, and closets require coat hooks.

Two types of locks are used in residential construction (see Fig. 12-29). The mortise lock is fitted into the edge of the door. Its mechanism is usually complicated, and therefore it is the type of lock used for greater security. Modern mortise locks can be fitted with a variety of lock cylinders and turn buttons.

Tubular locks are fitted through a hole bored through the door. They contain the locking mechanism within the door knobs. Because of their simple design they are less expensive than mortise locks, and they are easier to install.

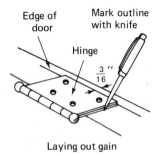

Laying out gain

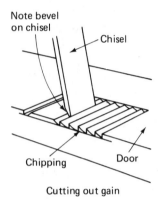

Cutting out gain

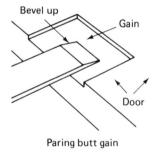

Paring butt gain

Fig. 12-28 Gaining for a Butt Hinge

Most locks used in residential construction are of tubular design. Locks from different manufacturers have different features, but all manufacturers produce locks which fit into one of three categories. These are passage locks, privacy locks, and entry locks.

Passage locks are simple locks used to control the door. They consist of door knobs and the necessary mechanism to operate the latch but have no locking device. This type of lock is used on closet doors, hall doors, bedroom doors, and other doors where no locking is required.

Privacy locks are similar to passage locks but have a simple locking device added to the lock mechanism. This type of lock is used on bedroom and bathroom doors.

Entry locks are provided with a button for locking on the inside and a key cylinder on the outside. This lock is used on entry doors and any other door which requires security.

The installation procedure for all types of tubular locks is similar but will vary slightly among locks made by different manufacturers. It is important, therefore, to follow instructions provided by the lock manufacturer.

Door Lock Installation   Holes for the tubular locks may be located by using the template provided or by using a jig boring template. After the holes have been bored for the knob assembly and the strike assembly the edge of the door must be mortised to receive the latch face plate (see Fig. 12-30).

Mortise locks require a large mortise in the edge of the door. This mortise may be cut with a special mortising jig, or it may be made by boring several holes with the aid of a template layout. The space between the holes bored into the edge of the door is pared clean with a chisel.

STAIRS

Stairs used in residences are usually made in a mill and delivered to the job for fitting and installation. They are usually built of wood and may be constructed in a variety of shapes. The most common stair is a straight flight (see Fig. 12-31). In any flight of stairs the height and width of each step must be uniform from end to end. The steps should be skid proof and should not contain any obstructions. The stair should be provided with a sturdy handrail which is smooth and free of slivers.

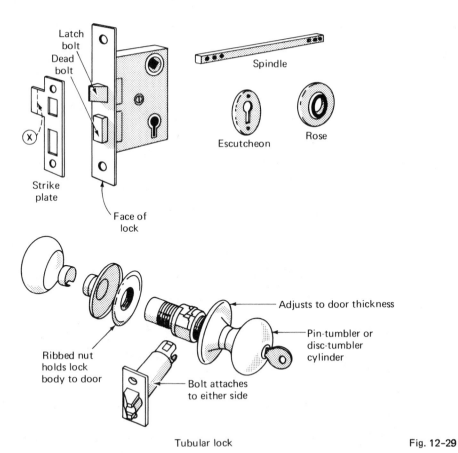

Latch bolt

Dead bolt

Strike plate

Face of lock

Spindle

Escutcheon

Rose

Adjusts to door thickness

Pin-tumbler or disc-tumbler cylinder

Bolt attaches to either side

Ribbed nut holds lock body to door

Tubular lock

Fig. 12-29   Locks

Fig. 12-30   Lock Installation

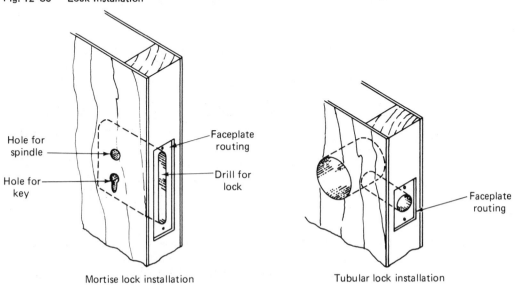

Hole for spindle

Hole for key

Faceplate routing

Drill for lock

Mortise lock installation

Faceplate routing

Tubular lock installation

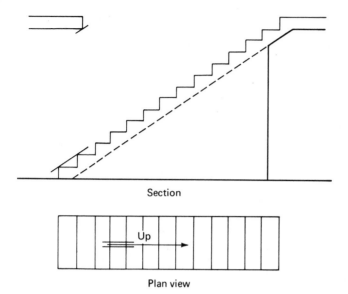

Section

Plan view

Up

Fig. 12-31   Straight Flight Stair

The shape or design of the stair changes depending on the available space for the stairway. (see Fig. 12-32). The quarter-turn stair uses a landing at the point where the stair makes a 90° turn, the half-turn stair is used when a 180° turn is required, and a winding stair is used if there is an extreme shortage of space for the stairway. The winding stair utilizes winding treads to make the turn in place of a landing. Winding stairs are to be avoided whenever possible because the narrow tread at the converging end makes this stair dangerous, especially for elderly and very young persons.

All stairs should have adequate headroom (see Fig. 12-33). For basement stairs the minimum headroom is usually 6'4", but other stairs usually require 6'8" headroom. Sufficient headroom not only keeps people from bumping their heads on the underside of the floor construction, but also provides adequate clearance for moving furniture or other items up or down the stairs.

In stair layout the total vertical distance from finish floor to finish floor is called total rise. The height of each step is called unit rise, and usually does not exceed 8". The board that is placed at the back of each step is called a riser. Not all stairs are fitted with riser boards. The mathematical width of the step is called unit run. A nosing is added to the unit run and the combined distance is called tread width (see Fig. 12-23). The unit rise and the unit run in any flight of stairs must be uniform

Fig. 12-32   Types of Stairs

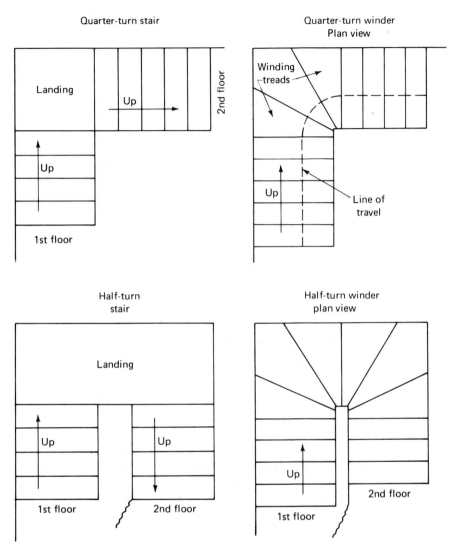

Quarter-turn stair

Quarter-turn winder
Plan view

Half-turn
stair

Half-turn winder
plan view

from top to bottom. The sum of the unit rise and the unit run should be between 17″ and 18″.

The opening in the floor through which the stair passes is called the stairwell or wellhole. The length of the stairwell governs the actual amount of headroom on the stair. Short stairwells reduce headroom, but long stairwells, although they increase headroom, use more floor space.

Fig. 12-33   Stair Terminology

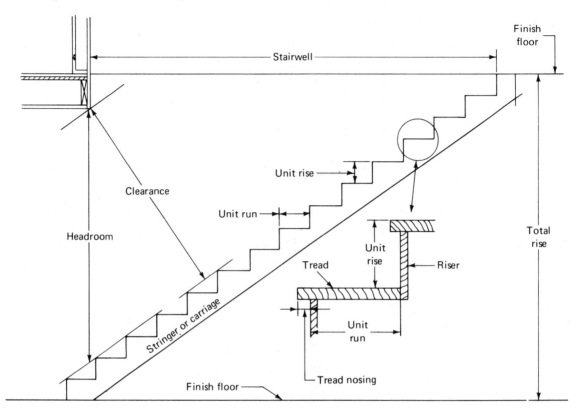

The part of the stair that supports the treads and risers is called a stringer or carriage. It may be made from a variety of materials. The cut-out carriage is often used for basement stairs, but the housed out stringer may be used on basement stairs and other stairs as well (see Fig. 12-34).

When installing stairs care should be exercised in cutting and fitting the various parts. All members should be securely nailed in place. When a housed out stringer is used it is advisable to glue and nail the wedges in place.

Fig. 12-34   Stair Stringer/Carriage

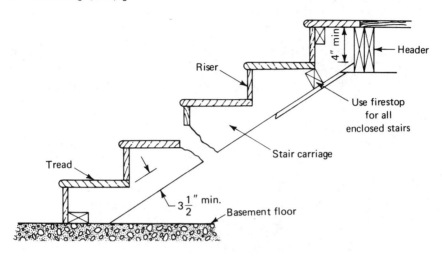

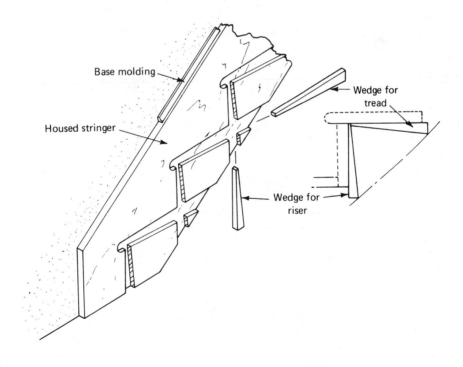

REVIEW QUESTIONS

1. How should a building be prepared to receive millwork?
2. What are door jambs?
3. Outline a procedure for installing door jambs.
4. What is the function of door casings?
5. When are door stops installed?
6. Outline a procedure for installing door casings.
7. Outline a procedure for installing window trim.
8. Describe the various types of base trim.
9. How are inside base corners fitted?
10. How are outside base corners fitted?
11. What is the standard height for clothes closet shelves?
12. How are linen closet shelves spaced?
13. Outline a procedure for installing kitchen cabinets.
14. Describe the two main types of door construction used in residential work.
15. How is the hand of a door determined?
16. Outline a procedure for hanging doors.
17. Outline a procedure for installing locks.
18. List and describe the main parts of a stair.

selected bibliography

American Plywood Association. *Guide to Plywood for Siding* APA, Tacoma, Washington, 1968.

——. *How to Buy and Specify Plywood* APA, Tacoma, Washington, 1966.

——. *Plywood Construction Guide for Residential Building* APA, Tacoma, Washington, 1968.

——. *Plywood Truss Designs* APA, Tacoma, Washington, 1964.

Anderson, L. O. *Wood Frame House Construction.* U.S. Government Printing Office, Washington, D.C., 1970.

——. *Guides to Improved Frame Walls for Houses.* U.S. Department of Agriculture, Forest Products Laboratory, Madison, Wisc., 1965.

Badzinski, Stanley, Jr. *Stair Layout.* American Technical Society, Chicago, Ill., 1971.

Durbahn, W. E., and Sundberg, E. W. *Fundamentals of Carpentry,* vol. 2. American Technical Society, Chicago, Ill., 1969.

Federal Housing Administration. *Minimum Property Standards.* U.S. Government Printing Office, Washington, D.C.

Huntington, W. C. *Building Construction, 4th ed. John Wiley &* Sons, Inc., New York, 1975.

Jones, Raymond P. *Framing, Sheathing and Insulation.* Delmar Publishers, Inc., Albany, N.Y., 1964.

National Forest Products Association. *Manual for House Framing.* National Forest Products Association, Washington, D.C., 1961.

——. *Design of Wood Structures for Permanence.* National Forest Products Association, Washington, D.C., 1961.

——. *Plank and Beam Framing.* National Forest Products Association, Washington, D.C., 1961.

——. *Span Tables for Joists and Rafters.* National Forest Products Association, Washington, D.C., 1970.

Smith, Ronald C. *Principles and Practices of Light Construction.* Prentice-Hall, Inc., Englewood Cliffs, N.J., 1970.

U.S. Department of Agriculture. *Wood Handbook.* U.S. Government Printing Office, Washington, D.C., 1974.

United States Gypsum Co. *Drywall Construction Handbook.* United States Gypsum Co., Chicago, Ill., 1971.

——. *Red Book of Lathing and Plastering.* United States Gypsum Co., Chicago, Ill., 1972.

Wilson, J. D., and Werner, S. O. *Simplified Roof Framing.* McGraw-Hill Book Company, New York, 1948.

index

index